IMPLEMENTING STANDARDIZED WORK

TRAINING AND AUDITING

The One-Day Expert Series

Series Editor
Alain Patchong

PUBLISHED

Implementing Standardized Work: Training and Auditing
Alain Patchong

Implementing Standardized Work: Measuring Operators' Performance
Alain Patchong

Implementing Standardized Work: Process Improvement
Alain Patchong

Implementing Standardized Work: Writing Standardized Work Forms
Alain Patchong

The One-Day Expert Series

IMPLEMENTING STANDARDIZED WORK

TRAINING AND AUDITING

Alain Patchong

CRC Press
Taylor & Francis Group
Boca Raton London New York

CRC Press is an imprint of the
Taylor & Francis Group, an **informa** business

A PRODUCTIVITY PRESS BOOK

CRC Press
Taylor & Francis Group
6000 Broken Sound Parkway NW, Suite 300
Boca Raton, FL 33487-2742

First issued in paperback 2019

© 2016 by Taylor & Francis Group, LLC
CRC Press is an imprint of Taylor & Francis Group, an Informa business

No claim to original U.S. Government works

ISBN-13: 978-1-4665-6363-6 (pbk)

This book contains information obtained from authentic and highly regarded sources. Reasonable efforts have been made to publish reliable data and information, but the author and publisher cannot assume responsibility for the validity of all materials or the consequences of their use. The authors and publishers have attempted to trace the copyright holders of all material reproduced in this publication and apologize to copyright holders if permission to publish in this form has not been obtained. If any copyright material has not been acknowledged please write and let us know so we may rectify in any future reprint.

Except as permitted under U.S. Copyright Law, no part of this book may be reprinted, reproduced, transmitted, or utilized in any form by any electronic, mechanical, or other means, now known or hereafter invented, including photocopying, microfilming, and recording, or in any information storage or retrieval system, without written permission from the publishers.

For permission to photocopy or use material electronically from this work, please access www.copyright.com (http://www.copyright.com/) or contact the Copyright Clearance Center, Inc. (CCC), 222 Rosewood Drive, Danvers, MA 01923, 978-750-8400. CCC is a not-for-profit organization that provides licenses and registration for a variety of users. For organizations that have been granted a photocopy license by the CCC, a separate system of payment has been arranged.

Trademark Notice: Product or corporate names may be trademarks or registered trademarks, and are used only for identification and explanation without intent to infringe.

Visit the Taylor & Francis Web site at
http://www.taylorandfrancis.com

and the CRC Press Web site at
http://www.crcpress.com

Contents

Preface vii
Acknowledgments ix

Chapter 1 Introduction 1

Chapter 2 The End of Industry 5

Chapter 3 Training Day 11

Chapter 4 Preparation for Training 23

Chapter 5 The Four-Step Method 37

Chapter 6 Practicing Training: The Second Chance to Get It
Right 47

Chapter 7 Introducing Auditing 57

Chapter 8 Key Points in an Auditing Document 67

Chapter 9 Sustaining and Auditing in Daily Management 73

Chapter 10 The End 79

Index 85
Author 89

Preface

The One-Day Expert series presents subjects in the simplest way, while maintaining the substance of the matter. This series allows anyone to acquire quick expertise in a subject in less than a day. That means reading the book, understanding the practical description given in the book, and applying it right away, in only one day. To focus on the quintessential knowledge, each The One-Day Expert book addresses only one topic and presents it through a streamlined, simple, and narrative story. Clear and simple examples are used throughout each book to ease understanding, and thereafter, application of the subject.

Acknowledgments

The One-Day Expert series is the direct consequence of my previous work at Goodyear. I owe thanks to several of my former colleagues who provided me with valuable remarks and comments.

I am thankful to Dariusz Przybyslawski and Mike Kipe, who were part of the team I formed to deploy Standardized Work in Goodyear plants. Dariusz's help was instrumental in structuring and tuning The One-Day Expert series.

I am very grateful to two other former colleagues at Goodyear who, at a very early stage, believed in Standardized Work as presented in the dedicated One-Day Expert series and gave me the opportunity to try it on the shop floor: François Delé and Markus Wachter.

I am obliged to Xavier Oliveira, Philip Robinson, and Gaël Coudeyras-Charlaté, who read drafts and offered valuable suggestions for improvement. I am also very thankful to the editorial staff of Taylor & Francis, especially Syed Mohamad Shajahan of Techset Composition, for their wonderful work in improving the readability of the initial text.

I express my gratitude to all my former colleagues at Faurecia who worked with me to test ideas and actions, and critique or support my thoughts.

Finally, and most especially, I give my special thanks to my wife Patricia, my son Elykia, and my daughter Anya for their unrelenting support and patience.

1

Introduction

In The One-Day Expert series dedicated to Standardized Work, Thomas, a young, high-potential plant manager in an industrial group, is reassigned to a plant that is losing money. Previous plant managers have tried several initiatives, with, to say the least, limited results. Thomas's urgent mission, which sounds like the EMEA,* senior management's last card, is very simple: turn the plant around. The morale in the plant is very low; the staff is equally pessimistic about the plant's future and distrustful of senior management. Time is running out; company's headquarters needs concrete results and has become impatient. To face these challenges, Thomas has decided to use Standardized Work deployment to achieve quick and visible results while rebuilding a real team. To this end, he has requested the support of Daniel Smith, the industrial engineering manager for EMEA. Daniel has been with the company for only a couple of years, after previous experience in the automotive industry. Building on his previous experience, he recently designed and launched a Standardized Work initiative and is looking to prove the real power of Standardized Work by deploying it in several plants.

The series of books dedicated to Standardized Work implementation recounts, step by step, Thomas's deployment of Standardized Work with Daniel's support. The first book of this series shows the initial steps Thomas took to assess the plant's current situation through measurement of operator's performance. The second book recounts the next steps of this assessment, which consist of writing Standardized Work forms to help see both variability and waste. It also sheds a light on a dire industrial engineering community squashed by a ubiquitous Excellence Systems† organization, and locked in a perilous and unrelenting fight for relevance, if not

---

* EMEA stands for Europe, Middle East, and Africa.
† Excellence System is the local Lean implementation organization.

survival. In the third book, dedicated to Process Improvement, Thomas opens a new "front line" in his quest to turn around the plant as he tries a new type of relationship with the labor union based on mutual trust and constructive partnership. Simultaneously, he negotiates a competitiveness plan, a sort of "give-and-take," wherein the management commits to keep the plant running and in return requests a compensation cost reduction. The third book also uncovers how he continues to push relentlessly for implementation of Standardized Work, and uses quick wins provided by improvement actions to raise the morale of his troops, even amid heated pressure from EMEA regional offices. More precisely, the third book shows how to use Standardized Work documents to generate and implement fast, simple improvements that produce quick wins and keep energy high while implementing all of the Standardized Work steps.

This is the fourth book of the series on Standardized Work deployment.* It is dedicated to methods and tools that are used to train operators, and thereafter check their work against the standard described in Standardized Work forms. It also discusses how to introduce those tools in the most effective way. These are key steps to sustain the implementation of Standardized Work. As stated in the first book of this series on Implementing Standardized Work,† for standards to be really sustainable and beneficial, they have to be implemented as part of a systematic approach including many other activities. Two of those activities are training and auditing. They are the last links that close the loop of Standardized Work implementation and make sense of all the implementation tasks enacted thus far, including capturing the initial state and improving the process. As a matter of fact, even the smartest improved process described on documents makes little sense if it is not actually applied on the shop floor. Making it happen at the workstation is the key to the success of the whole Standardized Work implementation. Very prosaically, this is where you start putting money in your pocket. Everything done before simply becomes a huge waste if you cannot apply and, most importantly, sustain the application on the shop floor. As a reminder, the

---

* As a reminder, the previous three books deal with the initial steps of Standardized Work deployment, which consist of capturing the current state and improving the process. The first book addresses operator performance assessment. The second one is dedicated to tools that help capture the current state of the process itself: the Process Analysis Chart, Standardized Work Combination Table, Standardized Work Chart, and Operator Work Instructions. The third book shows how to use the forms presented in the second book to improve the process. It also introduces simple and practical tools to generate quick improvements.
† *Implementing Standardized Work: Measuring Operators' Performance.*

benefits of Standardized Work include: improved safety, better quality, higher productivity, reduced costs, and increased team morale; all of these are the foundation of operational excellence. The current book is about how to effectively translate the aforementioned benefits into reality in a sustainable way; it encompasses two stages: training and auditing.

When it comes to training, the challenge is twofold: preparing for training and training itself.

The first point, preparing for training, is mainly about how to structure and present the newly improved way of working, as captured in Standardized Work forms, to facilitate the training. To this end, a new form is introduced: the *Job Breakdown Sheet*. The Job Breakdown Sheet is essentially a working document that includes most of the content from the Standardized Work forms described in the second book of the series.[*] Contrary to previous documents mentioned in this series on Standardized Work, the main focus of the Job Breakdown Sheet is its effectiveness, rather than its format. It contains key points that are not arranged in the same way as in the Operator Work Instructions.[†] It also gives the "why" for actions, as it is proven that people are more likely to adhere to instructions when they know the reason. Also, like the Operator Work Instructions, it provides illustrations to key points.

The second point, the training itself, is chiefly about the methodology used to deliver the training. The methodology presented in this book is not only simple to implement, but also, and most importantly, it has demonstrated tremendous success over the years. Practitioners know this method as the *four-step method*, from Training within Industry (TWI).[‡]

This book shares a slightly customized version of the Job Breakdown Sheet and the four-step method. It is worth underscoring that this is not a book for training experts, nor it is on TWI. It provides simple actionable

---

[*] The second book is titled *Implementing Standardized Work: Writing Standardized Work Forms*.

[†] Details about these forms can be found in the second book of the series, *Implementing Standardized Work: Writing Standardized Work Forms*.

[‡] Here is what Wikipedia says about Training within Industry: "The Training within Industry service was created by the United States Department of War, running from 1940 to 1945 within the War Manpower Commission. The purpose was to provide consulting services to war-related industries whose personnel were being conscripted into the U.S. Army at the same time the War Department was issuing orders for additional materiel. It was apparent that the shortage of trained and skilled personnel at precisely the time they were needed most would impose a hardship on those industries, and that only improved methods of job training would address the shortfall. By the end of World War II, over 1.6 million workers in over 16,500 plants had received a certification." (http://en.wikipedia.org/wiki/Training_Within_Industry, accessed 08/21/2014).

tools that will help the reader perform quick and effective operator training, which is a key step of Standardized Work deployment.

The second aspect of sustainable Standardized Work implementation is auditing. This is where application on the shop floor is checked against the results expected from the previously agreed methodology. Subsequently, the reasons for any gaps will be addressed, to close the loop and continuously improve the implementation of Standardized Work. This step is the "C" and, most importantly the "A" of the PDCA,* which is the heart of any implementation initiative. By experience, only a small number of people will reach the training step, and among those, the few people who get to the auditing stage tend to forget the "A" of PDCA. It should be underscored that auditing is very delicate to implement, as it tends to create resentment among operators, some of whom might perceive it as an act of spying and as a lack of confidence. Therefore, it has to be introduced carefully. This book gives some key points to ensure the successful establishment of auditing processes. The other point about auditing is the organization: "who does what, and when?" There is a role to be played by everyone in the plant to achieve sustainable success. Everyone's role, from the plant manager to the operator, is described and illustrated by simple examples in this book.

---

* PDCA stands for Plan, Do, Check, and Act. As mentioned in the previous book, as with any deployment, the Standardized Work rollout should be implemented following the PDCA approach.

2
The End of Industry

That night, as on previous ones, although Thomas arrived in his hotel room exhausted, sleep would not come easily. The temperature had started to drop and the wind was kicking up. Driving from the plant to the hotel required a lot of effort. "As Daniel explained today, I was in autopilot mode," he said to himself. While waiting to fall asleep, his mind wandered. He thought about the training day, how dense it was, and was quite happy that it ended with some results that he celebrated to keep the energy high in his team. He thought that his role as a leader was important, but he wanted to develop more leadership around him to ensure everlasting results. He strongly believed that real success would be powered by teamwork. In this sense, he was not a big fan of industry leaders strutting in the media, telling everyone who wants to hear that it's all about their genius. In his mind, the person who best illustrates this "me first" account of success was Jack Welsh.* Thomas would clearly agree with Jim Collins† that humility is an essential part of a great leader. He lamented that fewer and fewer senior managers around him showed this character trait. To his disappointment, he noticed that most people perceived humility as a sign of weakness or incompetence, and therefore a disqualifier for leadership roles. The conventional wisdom was that "you need to show your teeth and play lot of politics," Thomas conceded.

As sleep was long to come, Thomas's mind ventured deeper and deeper into more philosophical questions. He was reminded of an article he read this weekend while commuting back home, in which the author made a convincing case about the decadence of industry in France. The title, which read: "France: The End of the Industry," sounded like a book by Micheline

---

* Jack Welch was chairman and CEO of General Electric between 1981 and 2001.
† In Collins, Jim. 2001. Level 5 leadership: The triumph of humility and fierce resolve. *Harvard Business Review*, January: 66–76.

Maynard[*] he read a few years ago: *The End of Detroit*. In the article, the author, an expert on industry, made a very convincing case about what was going wrong in French industry, which over the years had become one of the least competitive among developed manufacturing countries.[†] The article laid out all shareholders' responsibilities: the government and politicians, industry leaders, and worker unions. As the author explained elegantly in his article, "This gang of three awkwardly teamed up to tear down French industry." The author faulted the government and politicians in general, first for not being able to see the problem early enough. He then asserted, "Once they recognized the problem, they made a series of wrong decisions." Thomas agreed with the author when, while dismissing the government's current approach, he claimed, "When it comes to competitiveness, I have noticed that it is often approached from a macroeconomic vantage point, thereby overlooking other key aspects of the subject, like technological, industrial, and supply chain factors—to say the least." Although Thomas had been very busy, he did have a few discussions with Eric, the Excellence System manager, who shared this point of view as well. Eric repeatedly told him that every time there was a number signaling a productivity gap between Germany and France in the news, the same people would show up to make the same case, "France needs to reduce labor costs." He explained that their point was that a French worker costs 10% more than one in Germany, which they believe to be the root cause of the loss of competitiveness. The author of the article pointed out that even if the labor cost was a real factor, it was not the most critical one. He complained that "it is unfortunate that the government has been following that idea, and pouring huge amounts of money into overhead cost reduction schemes that have benefitted all companies: big and small, the wealthiest and the neediest indiscriminately." Then he added, "Wealthy companies ended up grabbing millions they didn't need, which would have been of the greatest help to a large number of small companies that are in bad shape and scramble every day to keep their half-drown heads above water." The key, according to the author, was manufacturing excellence, "which was the heart of German company success." The article also accused some industry leaders of a lack of deep

---

[*] See Maynard, Micheline. 2004. The end of Detroit: How the big three lost their grip on the American car market. *Crown Business*, September: 368 pages.
[†] *The Shifting Economics of Global Manufacturing: How Cost Competitiveness Is Changing Worldwide* by Harold L. Sirkin, Michael Zinser, and Justin Rose (August 19, 2014), available at http://www.bcgperspectives.com, accessed on April 26, 2015.

business expertise. In times of fat cows, the author argued, you can play dice and win; there are a lot of winning numbers. It is therefore easy for general managers who have not mastered their business to thrive. As he put it, they need "to not do stupid stuff." "French industry leaders," he argued, "are less technical than German or Japanese leaders. The results are visible, especially in the automotive industry." The author wrote, "In today's highly competitive environment, industry, which has already harvested the low-hanging fruits, cannot rely anymore on single-minded or one-size-fits-all tools. Experts with holistic views and deep insights are needed." Another point that the article underscored concerned the way that French industry leaders were always eager to ship assembly activities to low-wage countries. "Many decisions were made based on very 'optimistic' calculations that did not take into account all related logistic costs and risks." They neglected a lot of hidden costs and overlooked inherent risks. Outsourcing has become the new mantra: "You ought to do it, or you had better come with strong justifications." The problem, the author asserted, was that most of those technically weak leaders could not challenge the status quo or engage themselves in any out-of-the-box thinking. Not only were they not in the best position to do so, but also it was simply too much of a risk for their career. As a result, French industry outsourced massively as compared to the more "pick-and-choose" approach of German industry. The Germans did outsource, the article said, but they chose to ship out only manually intensive or low-technical content tasks, while keeping the finest activities home. The author took the example of Volkswagen, the German automaker, which despite selling 10.1 million cars worldwide with 118 plants in 31 countries, has a workforce that is still 45% German.

Thomas agreed with the article's writer that in tough times, knowledgeable people—experts fluent in business who can make educated decisions—should be put in charge, as the confidence interval to hit the target narrows. People with only general manager's skills, with disdain for technique and love of politics, who are so common in French senior management, are riskier bets. By being more in focus on their own careers, just like politicians, French industry leaders, the paper insisted, have minimized or even abandoned all decisions that might hinder their career. This trend has meant minimum innovation and thereby a narrow growth horizon. Nobody has tried to grow the cake, therefore everyone has ended up fighting for a smaller portion. Lesser portions for an increasing number of people transformed the industry into a jungle. The author recounted how over the years he has seen work conditions toughened. "People befriended easily at their workplaces when I started my career as a young engineer," the author wrote. "Today I see lot of violence in the relations between people in workplaces, as everyone fights for his or her relevance, if not survival. Margins are getting lower, especially for the auto-component sector. As a result, the pressure on employees to deliver is high." As an interviewed employee put it in the article, "Over time I have noticed that more than talent or any other quality, you need the 'skin of a rhinoceros' to survive in our company." Another employee interviewed in the article recounted how this fear of risk pushed a big company to renounce inclusivity. "The HR staff were very comfortable hiring the same kind of people, with the same background, who had graduated from a selected panel of universities; they were not ready to take any chance on being inclusive. It's amazingly surprising in my company, where everything should be about creativity and innovation. I happen to believe that we can only flourish when we embrace people's differences." The author called this phenomenon the "cloning spiral," which he said was putting French industry and the political landscape at risk.

The part of the article on worker unions was equally tough with sound arguments. The author contrasted the situation to the more constructive German atmosphere, in which it was possible for a former head of IG Metal[*] to take the helm of the world's second-largest automaker. "In France, they have been holding the shovel that dug the hole," the author asserted. The article described them as very rigid, defiant, and in most circumstances opposing any new reform ideas coming from industry leaders. Not only did they not look for compromise, but very often they also encouraged their members to take radical actions ranging from frequent strikes to asset destruction. The results of this permanent fight prevented stakeholders from coming together and solving problems, which did no good for a declining industry. Unions systematically took radical, even extremist paths, ready to fight until complete surrender of the other side. We saw weeks, even months of strikes costing millions to corporations. The author insisted that all this lack of cooperation could still have a limited impact if the economy was going well, but in tough times, it was a

---

[*] IG Metal is the powerful metalworkers' trade union in Germany. His former head, Berthold Huber, stepped down in 2013 to become deputy chairman of Volkswagen's supervisory board and was appointed interim chairman after the departure of Ferdinan Piech, the previous holder of the position.

recipe for disaster. He then concluded that without changes, French industry would go from a death spiral to complete extinction.

Thomas had some thoughts of his own, "It's obvious that things are not going so well in this country." He had been spending lot of time in France, and as a person with an international background but mostly German culture, it was hard to understand some French habits. This was not a subject of complaint. On the contrary, as someone who wanted to learn more, his new assignment in France was a great opportunity to confront himself with differences and enrich his background. He glanced at his watch, jumped, scratched his head and murmured, "It's already 1 a.m., I seriously need to sleep now if I want to be productive for the last day. Let's put everything away. As the saying goes, tomorrow is another day."

Friday was, in effect, the last day of the weeklong training. As far as he could remember, it would be about training and auditing, the last steps of the Standardized Work deployment. Thomas turned off the lights, and within a few minutes he fell asleep, subdued by the fatigue of long and demanding days of work.

3

Training Day

Like on previous days, Daniel arrived quite early. As always, he needed to complete some preparatory tasks before the group showed up. The ritual was now pretty standardized. First, he wrote down key messages and drawings from the previous day; these would be used to check the retention level of the previous day's training. Second, he crafted and displayed the paperboards regarding training and auditing Standardized Work. As today was the last day, he also added a few charts for the overall conclusion of the training he called "Making sense of all."

Friday was not as bright of a day as one would expect for this time of year. As a local trainee characterized it, "The weather was in between the kind we are used to in this area." No one around would complain about the weather. There was a common saying, which elicited very much the spirit of the people here. In effect, everyone here would tell you: "The sun we do not have outside, is in the heart." No surprise that despite the gloomy time outside, there were a lot of bright faces in the room on this last day of the training. Among those radiant faces was Sarah's.

Sarah, the HR manager, was the only female manager in the plant. Here, as in most companies in France and around the world, there were very few women in industry. There were even fewer in upper management. As elsewhere, the few you would see were mostly in the position of communication manager or human resources manager. Although she held an HR position, Sarah actually attended one of the best "Grandes Ecoles" in France. As she would often recount, Sarah loved the shop floor of the plant. She wanted to be involved in manufacturing tasks. This is actually why she joined her current company, which offered her a job in production right after graduation. When she took the job, she was a very enthusiastic young engineer running around all day long to ensure the required production. She adored the high level of adrenaline related to her job. At that time, she had a young boss who was very enthusiastic as well. "He was a very supportive guy; every day I came to work I was thrilled to be part of his team," she would insist. "I felt like we were changing the world." Her young boss quickly got a promotion and was replaced by an older manager. "A kind of old-style, man's manager," she would specify. The relations between the two worsened very quickly, as the newcomer would from time to time adopt abusive and derogatory language. Sarah started suffering from misogynic acts from her colleagues. As she explained, "I guess those acts had always been there, but I had not seen them because I was so involved in the daily business; after the arrival of my new boss, I started noticing and resenting that, and I became very intolerant of everything that I considered macho. I had had enough of his management style, infused with micro management; I needed a change. I was not feeling well anymore. Therefore, I reached out to my boss's superior. I had easy access to him; we went to the same school of engineering. He offered to move me to the HR department, where I would be in charge of training. He thought that my shopfloor experience would be a very valuable asset. As he put it, 'The best HR managers are the ones with business experience.'" In a move that resembled a self-realizing prophesy, Sarah, just like many women, ended up in the HR department. Thanks to her hard work, she was promoted very quickly. Besides being the HR manager of the plant, she quickly became a key person who had her say in important topics. Paradoxically, by joining the HR department, she gained the kind of credibility and respect that she had unsuccessfully fought so hard to acquire in manufacturing. Sarah, a former reader of fantasy books, would willingly paraphrase Garth Nix's quote in *Sabriel** and admit, "I guess when it comes to my career, the 'path chose the walker,' but I am very happy to be where I am today. I really love being in charge of human resources in the plant. It gives me a unique vantage point that I could not imagine." Not everything she did in her job made her happy. In times of uncertainty the company would need to "let people go"—the phrase she used for "firing people." Once it was decided, often not by her, she was the executioner. This was quite in contrast to the situation she found when she became the HR manager. At that time, the plant was flourishing. She still remembers the sparkling light in candidates' eyes when they signed their contracts; especially new graduates who were signing onto their first jobs. "I got fired-up all day long after signing a new graduate," she recounted. Those days had gone, and she now needed

---

* "Does the walker choose the path, or the path the walker?" in *Sabriel* by Garth Nix.

to find the rationale to "keep the fire alive." The best formula she came up with was the French version of a book title. As she explained to Daniel, "In everything I do, 'there are God's deeds and a part from the Devil.'"* She had convinced herself that even if she didn't like to do certain things, she needed to do them for the good of the whole plant, just like the doctor in *Cider House Rules*. Sometimes she would tell, with a lot of emotion, how she moved heaven and earth to avoid firing the first person she had to fire. "The volume had been down for a while. We were running out of cash, and needed support from the regional office. They took this opportunity to focus on our cost. What we had been fearing quickly come to the table. They wanted us to cut costs, and the labor was the obvious burden on our cost structure. From that point we had no other choice but to let go people we had no complaint against."

Daniel quickly discussed Thursday's work. He asked participants what their takeaways from the training were. Most trainees mentioned how easy it was to use the Standardized Work documents to identify quick improvements. "Everything was so visual," Sarah, the HR manager, said. "Simply by looking at the Standardized Work Chart and the Standardized Work combination table, the waste at the workstation appeared very clearly," she added. Eric, the Excellence System manager, raised his hand to add more comments. "I liked the documents[†] you showed regarding workstation assessment; I find the process quite simple and very visual. My feeling is that I would have no problem to use it, even as an autonomous tool." Flattered by such feedback, Daniel effusively thanked Eric. It was clear that he was very proud of this result from training he invented from scratch only a few years ago. The time was right to move to the training day's agenda. As always, he went back to the chart he explained on day one (Figure 3.1):

> As you are all aware, we have now covered the first and second blocks of this chart, which respectively consist of capturing the current state and improving the process. Today, which is the last day, we will be addressing the last two blocks. I would like to detail a little bit of what today's training will be about. There are essentially two parts: training and auditing. When I say training, it's about how to effectively train operators to use the best method we have worked so hard to describe. Let me insist; this

---

* Author's translation to English of *L'Œuvre de Dieu, la part du Diable* whose original in English is *The Cider House Rules*, a novel by John Irving.
† For more details, see the third book of the series: *Implementing Standardized Work: Process Improvement*.

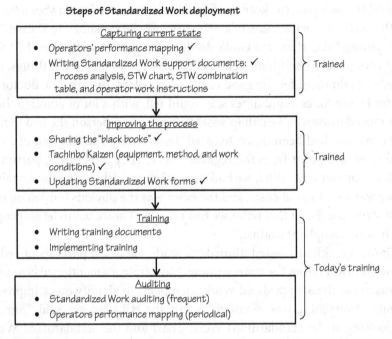

FIGURE 3.1
Different steps of standardized work implementation.

is very important: do not expect an operator to work as described on the Standardized Work forms if she or he is not trained. Remember how much effort we have put into our work so far. Also think about the potential savings we estimated. I need to tell you that I am always puzzled when people do so much work to find the best way to improve a process, and afterward leave its application to chance. It is just unbelievable, and I can tell you, this happens very often. Folks tend to think that the most difficult thing is to find the best method for a job, and once it is described, the rest is a "piece of cake." It's actually the contrary. While it is very important to identify good practices, at the end of the day, the most important thing is the bottom line. I mean, it is preferable to have a few good practices that are actually applied on the shop floor than great ones that are not followed. Well, guess what? "Training" is a key process in the application and it should not be considered easy stuff, as I'm sure Sarah here can attest.

Sarah acquiesced, visibly happy with the validation of the importance of her duties by an "important" guest. So many times she had come across a new hire or a temporary worker hastily placed on a workstation to make sure he or she could become productive as soon as possible. Over the years she had had little success convincing people that training should be perceived not as a pure cost, but as an investment. She had a motto for that: "Pay now, or you will pay more later." She relentlessly explained to an unimpressed audience, "Workers need to be well trained upfront in order to reap all their efficiency, and ultimately obtain higher overall total output from them." She secretly hoped that Daniel would dive in, and counted on his authority to change her colleagues' minds. Sarah was distracted by her thoughts for a few seconds. When her attention returned to the room, she had the feeling that her hopes were being realized. Daniel had being drawing a chart on a paperboard (Figure 3.2). Daniel completed his graph, cleared his throat discretely, and proceeded. "I wanted to share this chart with you. The two curves represent the performance of two hypothetical workers. The first one, let's say Worker 1, is properly trained before starting his job. Of course he will spend some time away from the line producing nothing. For sure, this is a loss of production that some shortsighted manufacturing folks do not like. Now, when Worker 1 starts her or his job, she or he will reach top performance quickly. On the other hand what do we have with the other worker? Well, Worker 2 represents an unfortunately-too-common situation. Here, the worker is rushed to the line with little or no training. Presumably, this is a way to both save training cost and get some additional parts production. We see this very often when there is pressure to increase production to satisfy customer demand. Well, as you can see, the money you save early will be completely wiped out by the poor effectiveness of the worker over time." Daniel stopped a few seconds to check that the room was still following. Being the seasoned trainer he was, he knew

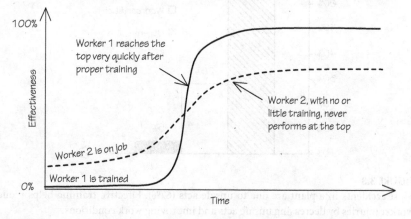

FIGURE 3.2
Over time, Worker 1, who receives a proper training before starting his job, will perform more effectively than Worker 2, who starts right away with little or no training.

that a good short story really makes knowledge stick. He then decided to tell a short story that happened to him in his adolescence. "Several years ago, along with a few friends, we decided to learn swimming. None of us had swimming experience. We took a subscription at a local pool. We were young, very competitive, and we liked challenges. Then we decided that we would learn swimming alone without a trainer. You know what? Well, we drank a few cups of pool water and we all succeeded. However, most of us, including me, did not swim so well. I learned the hard way, and to this day, I have never been able to improve my swimming. I still do not swim correctly. As a result, I really do not enjoy swimming. If I might summarize, poor or no training leads to the following consequences. First, it leads to wasted worker capacity, as top performance is difficult and often impossible to reach without proper training. Second, it can result in ingrained bad habits that are even more difficult to eliminate later on. Third, not knowing how to do her or his job can be a very stressful situation for a worker; this can lead to demotivation, quality problems, and, more importantly, safety incidents. Talking about safety, I am sure that you still remember the chart I shared with you yesterday,* which showed that 67% of safety issues are related to unsafe acts (Figure 3.3). Although some unsafe acts can be attrib-

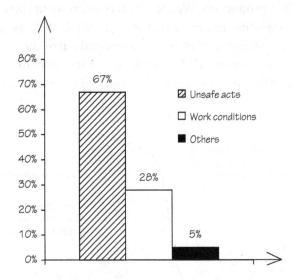

FIGURE 3.3
Most accidents in a plant are due to unsafe acts (67%). Effective training helps reduce worker injuries by decreasing unsafe acts and improving work conditions.

---

* For more details, see the third book of the series: *Implementing Standardized Work: Process Improvement*.

uted to bad behaviors, most of them are simply the result of ignorance due to poor training. When workers are well trained, you can eliminate a big chunk of the 67% of unsafe acts that appear in the chart."

> *Do not expect an operator to work as described on the Standardized Work forms if she or he is not trained.*

> *Pay now, or you will pay more later ... Workers need to be well trained upfront in order to reap all their efficiency, and ultimately obtain higher overall total output from them.*

Daniel paused for a few seconds. Thomas raised his hand and commented on how the importance of proper training had become clear to him after Daniel's explanation. He thanked Daniel, and encouraged Sarah to lean on him in the future if she encounters any difficulty carrying out her training duties. Sarah, a generally outspoken person, took the opportunity to insist on the importance of training in general. "Company leadership is always very keen to call people their most important assets, yet when it comes time to walk the talk, nobody is there. I even get the impression that the machines, which are maintained periodically, are better tended to than the people. People are mostly perceived as a cost that can be phased out in tough times to reach budgets. It is unfortunate that we do not understand that people should be seen not only as important asset, but also as appreciating assets, since their worth increases over time as they accumulate skills, knowledge, and experience. Machines are exactly on the opposite side. They depreciate over time." She also added that people should be a step ahead in planning HR needs the same way they are doing for volumes and production capacities to avoid being in the situation of Worker 2. "We spend plenty of time planning what should be our product in the coming months and even years, but we consistently fail to think about our need for people to support value creation for our customers, let alone succession planning. Getting the right people with the right competence does not happen over night; we need to plan ahead." Daniel took this opportunity to recount how companies with good practices like Toyota act, "When they set up a new plant, they usually hire workers several months ahead of the SOP,* send them to the closest regional training center, then to a nearby plant called the 'mother plant' for training on similar processes to make sure that they will be ready on day one. Just to give you a glimpse of the spirit at Toyota, one of the most common sayings in the company's leadership is 'We do not just build cars, we build people.' Please adopt this as your motto as well. You should understand that your key competitive advantage is people. Therefore, invest to develop and train them. Think about this, your competitors have access to the same suppliers as you, they can also hire the same young people coming out of universities as you, and they may even copy your products. Where do you think you get the opportunity to outperform them if it is not in the way you manufacture your products? Guess what, a big part of this 'way' is defined by employees. Therefore, they should be the subjects of special attention. They are our best weapon in game of outperforming competitors, and training helps to keep them appreciating over time, thereby solidifying our competitive advantage."

Poor or no training leads to wasted worker capacity, and it can result in ingrained bad habits that are even more difficult to eliminate later on. In addition, not knowing how to do her or his job can be a very stressful situation for a worker; this can lead to demotivation, quality problems, and more importantly, safety incidents.

Now that everyone seemed to understand the importance of training, Daniel felt that he had the best conditions to move on to the next step, which consisted of explaining that successful training needs content, an effective support document, and a qualified trainer. He made his point using a now-common image of a stool† (Figure 3.4). The content, he explained, was already in the previously identified key points. Therefore, the training module would focus on how to get ready for the training by writing an effective support document (Job Breakdown Sheet), and which method a trainer should use to achieve her or his goals efficiently (the four-step method). He underscored, "I should normally say preparation for training instead of writing of the Job Breakdown Sheet. The reason for this

---

* Start of production.
† Imagery previously used in the third book of the series on process improvement to present the types of wastes: Muda, Mura, and Muri.

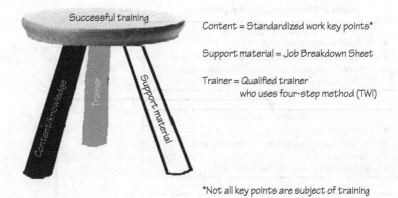

FIGURE 3.4
Successful training requires content, efficient support material, and a qualified trainer (who uses the four-step method).

shorthand is that although preparation for training includes other activities, as we will see later, the Job Breakdown Sheet is its main deliverable."

Daniel then moved to the next subject of the training: auditing. As previously, he wanted to make sure that everyone understood the importance of auditing. "How do you know that something you set out to do is actually done, if you do not actually check it? Auditing is simply common sense. I am appalled that some people might see it as optional. Frankly, managing a plan without auditing is like driving a car without cockpit information. Although the worst is not certain, the lack of information highly reduces the driver's chances of arriving safely. As I always explain, for any system to be deployed in a sustainable way, it has to follow the *Plan, Do, Check,* and *Act* steps. Auditing corresponds to the 'Check' but also includes the 'Act' as well." Daniel moved to a chart he had been using extensively (Figure 3.1). The chart showed the various steps of Standardized Work deployment. He added a feedback line and proceeded (Figure 3.5). "This feedback arrow symbolizes the 'A' or 'Act' of PDCA. This is the action you need to take whenever there is a gap between the observation and the defined standard. Obviously, you do not need to go systematically through all of the steps. For instance, if it appears that the gap between the application and the standard is due to the operator having been ill-trained, then you only need to retrain this worker. This gap might also be due to the fact that an operator, intentionally or not, is using a better method. Then the standardized documents should be updated and a training of all concerned workers conducted to maximize the benefit of the new method plant-wide and company-wide.

20 • *Implementing Standardized Work*

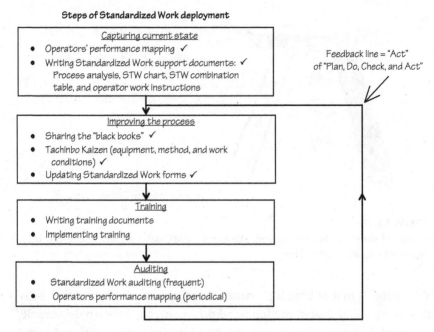

FIGURE 3.5
Auditing ensures continuous improvement.

Well, I won't list all possible cases. We will come back to this point when we focus on auditing later on. The key point I want you to remember here though is the feedback arrow I just drew, a symbol of auditing. It is the wheel that ensures continuous improvement. Without it, you should have improvement, but only a one-shot improvement, which happens when Standardized Work is established." Daniel stopped and asked if there were any questions. Nobody raised a hand, so he continued. "In today's training, we will be focusing on two key activities of the auditing steps. First, we will be discussing how to better introduce auditing in the plant. Let me say it very clearly, auditing must be introduced in a certain sensitive way to ensure its success. Second, another important point is what I would call the auditing routine, or the auditing organization. It consists of specifying everyone's role, from the operator to the plant manager. Who does what, and when? This organizational step is the key to achieving sustainability." Daniel punctuated and proceeded, "As you all know, this is our last day of training, we will take some time at the end to review the overall session, and to agree on the main takeaways and next steps. All right?"

Daniel stopped and offered a 5-min break. "When you return, we will talk about training. I have decided to make a slight change to my

> **Key take-aways**
>
> - Training and auditing are key steps that convert prior hard work into money in your pocket.
> - An operator should not be expected to work as described on the Standardized Work forms if he or she is not trained properly.
> - "Pay now, or you will pay more later" + training is an investment, not a cost.
> - Poor training leads to lasting wasted worker performance, demotivation, quality problems, and safety incidents.
> - Auditing corresponds to the "Check" but also includes the "Act" of the "Plan, Do, Check, and Act" method. It is the wheel that ensures continuous improvement.
> - Auditing is an efficient management tool.

FIGURE 3.6
Key take-aways from training and auditing introduction.

approach. I will ask you to try it first, and then we will use the trial to illustrate what you should be doing differently and how. Then we will try again while applying new learning to see the difference." Before the break, Daniel asked Thomas if he would summarize the content as he often does. Thomas went to the chart and wrote down some of the few points he retained (Figure 3.6).

> *How do you know that something you set out to do is actually done, if you do not actually check it?*

> *For any system to be deployed in a sustainable way, it has to follow the Plan, Do, Check, and Act steps. Auditing corresponds to the "Check" but also includes the "Act" as well.*

4

Preparation for Training

When the team returned from the break, Daniel wanted to address the question everybody had in mind, "What should be the content of the training?" First of all, he felt like he needed to give a quick reminder before discussing training.* "The T-shirt packing we have been using from the beginning to illustrate Standardized Work training is an operation cycle that has four major tasks to be done by a single person: (1) picking the T-shirt, (2) ironing the T-shirt, (3) folding the T-shirt, and (4) storing the T-shirt. As you remember, I initially asked you to split into four groups positioned according to the layout depicted on the flipchart over there (Figure 4.1).† I then asked every group to perform 40 cycles of each operation and compute their mode, which we also called standard time. You guys came up with times in this table (Figure 4.2). As part of the improvement process, we tried to identify reasons for the differences in times between teams in an effort to find best practices. We quickly agreed that the difference between groups for picking a T-shirt and storing a T-shirt, which by the way are very simple tasks, were mainly due to distance discrepancies. I mean distance from raw material bin to workstation, and distance from workstation to finished goods bin, respectively (Figure 4.2). We also agreed that variation in the third task was mainly the result of different folding methods. As you may note, ironing, a machine operation, is quite stable. The conclusion of this brief analysis is that the task with the biggest potential for training is folding. Now, let us step back a little bit to have a better view of the big picture. Remember when we listed all operations performed by each operator and obtained this tree (Figure 4.3).‡ 'Ancillary tasks' as well as 'Core tasks' must be considered

---

* More details related to Daniel summary can be found in the first three books of the series.
† Chart extracted from Improving Standardized Work: Writing Standardized Work Forms.
‡ Chart extracted from Improving Standardized Work: Writing Standardized Work Forms.

24 • Implementing Standardized Work

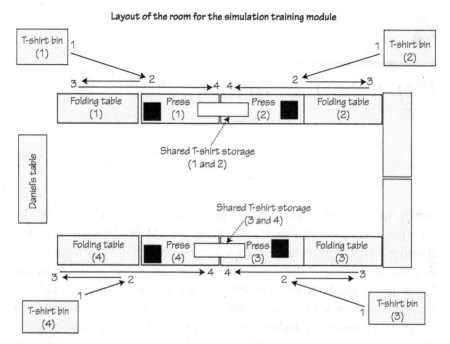

FIGURE 4.1
Layout showing the position of the four teams (not scaled); more details are available in *Implementing Standardized Work: Writing Standardized Work Forms*.

	Team 1	Team 2	Team 3	Team 4	Variation	Obvious reason(s) for variation
Picking the T-shirt	7	21	5	21	76%	Distance from raw material bin to station
Ironing the T-shirt	11	11	11	12	8%	Little difference. Press time very stable
Folding the T-shirt	11	24	19	28	61%	Operators have different folding methods
Storing the T-shirt	8	5	7	11	55%	Distance from finished goods bin to station
Cycle time	37	61	42	67		

FIGURE 4.2
Time variations between the four teams have different justifications. Variation in the first and the fourth task are due to distance, while variation in the third task is mainly the result of different folding methods—ironing, which is a machine operation, is very stable.

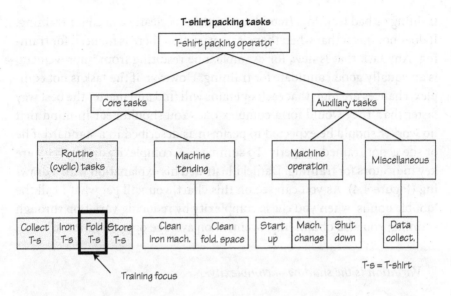

FIGURE 4.3
Breaking down T-shirt packing operation. Folding, which is the most variable task, is the focus of training.

for training. However, what I have said previously for improvement also applies to training: the priority must be given to cyclic tasks. The reason is the economies of scale principle. So we will focus on the T-shirt folding task as it holds the most promise for delivering training gains. Now, do not get me wrong, as we will see later, saving on an operation task will not always translate into saving on the operation as a whole."

At this point Daniel was interrupted by Eric, the plant Excellence System* manager. "Daniel, if you were to sum up, what would you say are the key characteristics of a task that are good candidates for training?"

Daniel cleared his voice and proceeded. "Well, Eric, that is an excellent question. We are looking, first of all, for cyclic complex tasks. When operators are not trained properly to do complex tasks, they naturally tend to operate very differently. I hope it is obvious to everyone; if not, stop me." Daniel paused a few seconds; nobody raised a hand. "Therefore, a good way to identify a complex task, besides seeing it, is to look at the variation that occurs from cycle to cycle with the same operator, and then from one operator to another one. Variation is the shadow of complexity. It could be real complexity or an operator-perceived complexity due to lack of

---

* Excellence System is company's customized version of the Lean implementation initiative.

training or bad training. In both cases, there is clearly a need for training. It does not mean that when there is no variation there is no need for training. Any task that is new, for example, one resulting from improvement, is an equally good candidate for training. However, if the task is not complex, chances are high that each operator will find and master the best way faster than they would for a complex one. You should bear in mind that no worker should be expected to perform as described in a standard if he or she is not trained properly. To summarize, complexity and novelty are key indicators for training. Daniel illustrated his explanation with a drawing (Figure 4.4). As you can see on this chart, you will get what I call the 'double bonus' when you chase complexity by reducing variation through training; you also obtain improvement on average operation times."

Variation is the shadow of complexity.

To summarize, complexity and novelty are key indicators for training.

Daniel paused and checked back with Eric to make sure that he was satisfied with the answer, then continued. "Now, as you may guess, we will be focusing the training on the folding task. As explained before the break,

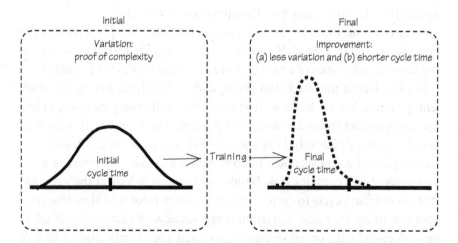

FIGURE 4.4
Variability is the shadow of complexity; training will help deal with complexity by both reducing variation and the average cycle time.

> Training process—role playing
> - We will focus on the complex operation only: T-shirt folding
> - Each team's member trains one person from another team to fold a T-shirt
> - Make sure performance criteria (quality, safety, and pace) are met
> - 10 min allocated for this exercise

FIGURE 4.5
Role playing tasks for group cross training.

I will be doing things slightly differently this time. I will confront you directly with the practice, and thereafter use the result to illustrate the key points of the training. Here is a summary of the tasks" (Figure 4.5). Daniel paired up participants and randomly asked one of them to train the other. Participants received no instruction whatsoever on how to train properly. Everyone needed to come up with his or her own method. At no surprise to Daniel, trainers proceeded differently. After the practice, Daniel convened the room for a thorough discussion and to share trainers' feedback. He structured the discussion around four questions: (1) What went well? (2) Did you achieve your target? If so, what helped to achieve the target? (3) What were barriers? and (4) What could be done better, or differently? Basically the overall feedback was that no one clearly succeeded in providing effective training. They felt like they did not have the right tools, or the method to accomplish their task. No method was imposed; everybody simply had to train the other person with his or her folding method. Everyone knew exactly what to achieve, and yet they all failed to convey the knowledge effectively. The discussion stoked the participants' appetite for learning, which naturally led Daniel to introduce a chart that showed how retention rates differ with tools and methodologies used (Figure 4.6). He called it "the learning hat."* The chart listed various teaching methods and offered an estimate of related retention rates. It appeared those methods could be split into two groups: passive ones and participatory ones. The latter holding the highest rates of retention. Daniel insisted that a good training session should have two stages. The objective of the first stage would be to bring the trainee up to "practice and feedback," which

---

* This is adapted from a chart commonly called the "Learning Pyramid." Although it seems completely intuitive, the author has not been able to verify the scientific base of this result, which is often mentioned as originating from the National Training Laboratories, Bethel, Maine. Other investigations led to similar conclusion—see Learning and Teaching Myths and Misconceptions, http://www.learningandteaching.info/learning/myths.htm, accessed on December 1, 2014.

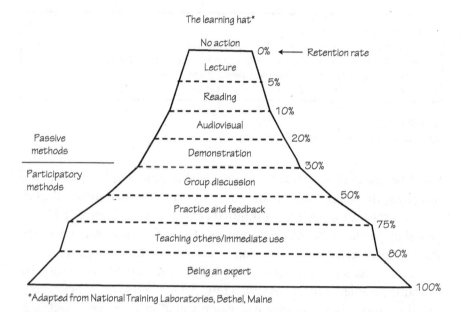

FIGURE 4.6
Different ways of learning and their retention rates.

corresponded to a 75% retention rate. "This is the level where you need to bring the worker after the training. The second step starts when the worker joins her or his workstation. Then, with the assistance of the team leader, the worker is expected to reach 80% retention quickly. The 100% retention rate will be achieved after a long time of practice, bringing her or him to the level of expert on task execution. Your goal is to acquire effective tools and methodologies that you will apply to bring the trainee to the maximum retention efficiently." Daniel spent a few minutes elaborating on the words "effective" and "efficient." He explained, "You should not only reach 100% retention rate, but this has to be done as quickly as possible to save the direct and indirect costs incurred by a low performing worker. In this sense, the TWI* tool and instruction method are quite aligned with the

---

* As a reminder, TWI stands for Training within Industry. According to Wikipedia, "The Training within Industry service was created by the United States Department of War, running from 1940 to 1945 within the War Manpower Commission. The purpose was to provide consulting services to war-related industries whose personnel were being conscripted into the U.S. Army at the same time the War Department was issuing orders for additional materiel. It was apparent that the shortage of trained and skilled personnel at precisely the time they were needed most would impose a hardship on those industries, and that only improved methods of job training would address the shortfall. By the end of World War II, over 1.6 million workers in over 16,500 plants had received a certification." (From http://en.wikipedia.org/wiki/Training_Within_Industry, accessed 08/21/2014.)

Learning Hat." Daniel also emphasized that to train properly they needed to prepare first. "If the preparation is skipped, the training will be missed. Obviously, none of you took a few seconds to build training plan, let alone any kind of support documents. You all rushed to start the training." He wanted to take this opportunity to underscore that the most important factors in a successful training are not the expertise in doing the job, but good preparation and delivery of the instruction. "Remember the story I told you the first day about one of our plants in the Netherlands where we found that the worst performer had actually been trained by the best performer. The management was then convinced that the problem came from the trainee. After analyzing the situation, we found that the worker had not been trained properly because his instructor, the best performer, lacked the basic qualification of a good trainer. I would like to take this opportunity to remind you of a famous quote from TWI, 'If the worker hasn't learned, the instructor hasn't taught.' It's clearly our responsibility to train our people, and when there is a problem, we should question the training before the worker's behavior. I am often amazed how quick managers are to point out worker behavior. It might seem convenient, but it is in no way constructive. Please do not get me wrong. There are, for sure, bad apples, but make sure first that the worker has been trained properly before jumping to 'behavior' fixing."

The most important factors in a successful training are not the expertise in doing the job, but good preparation and delivery of the instruction.

It's clearly our responsibility to train our people, and when there is a problem, we should question the training before the worker's behavior.

Daniel stopped and stared at the window. The weather outside was showing no signs of improvement. "No hope to see the sun today," he murmured to himself. He then turned to the room, cracked a joke about the weather and continued. "If you have no questions regarding what we have just discussed, I suggest we move to the Job Breakdown Sheet (JBS). This is the main deliverable of the preparation step. There is a sample in the pile of documents I just handed out (Figure 4.7). I think you all have a copy. Contrary to previous Standardized Work documents we have studied thus far, the main focus of the JBS is its effectiveness, rather than its format.

Job Breakdown Sheet

Description of the task:	Carcass building on ERP			List Common Key Points:
Parts (UPN, describe the parts):				
Tools & supplies required:				
Safety equipment required:				

Important Steps	Key Points		Reasons	Training Aid: (put hand sketches, diagrams, parts, or layouts here. Insert digital picture if available.)
WHAT? A logical segment of the operation that advances the work.	HOW? Things in important steps that will: 1. make or break the job 2. injure the worker 3. make the work easier		WHY? List the reasons for the key points	

#	Important Steps	Key Points		Reasons	
1	Start the machine	1. Stay away from the machine	✚	1. You might be hurt	[photo 1]
2	Pick up the bead	1. Make sure the bead comes alone	●	1. Beads often fuse	[photo 2]
		2. Pick the bead while the machine is running	◆	2. Needed to reach the cycle time	
3	Load the bead and re-start the machine	1. Keep the splice setting at 12 o'clock	◆	2. Guarantees tire uniformity	[photo 3]
		2. Place the bead around the bead-setter up-down carefully	✚	2. To avoid pinching your fingers	
4	Splice the ply	1. Ensure splice overlaps on 3 to 6 cords	◆	1. More than 6 cords → uniformity problems, less than 3 → splice opens leading to scrapped tire	[photo 4]

| Key Point reminders: | △ Critical check or inspection | △ Quantity Check | ✚ Could injure the person | ● Makes the job easier | Owner of this document: | Page: of | Rev: Date: |

FIGURE 4.7
Example of a Job Breakdown Sheet.

> Key characteristics of the Job Breakdown Sheet (JBS)
>
> A. Tasks are broken into small pieces
> B. Include key points (safety, quality, cost, and productivity)
> C. For each key point, there is an explanation
> D. It is a free-format document
> E. Include pictures/drawings, or audiovisuals that can support training

FIGURE 4.8
Main characteristics of a Job Breakdown Sheet.

On the chart over there are five points that sum up the key characteristics of a JBS (Figure 4.8). I will read them one by one, and use the sample you have in your hands to illustrate. First of all, this document looks like the Operator Work Instruction that you see here (Figure 4.9),[*] but they are completely different. I will illustrate the differences as we go through. The first point is on the content of the JBS (Figure 4.8). As indicated in the name, the 'Job Breakdown Sheet' breaks down the overall operation into small pieces. The rationale for this is that according to experts, the best way to teach people to perform an operation is to break it down into small chunks first, then teach it piece by piece, as the whole operation is being built up. For this reason, the JBS contains operations broken in small pieces. I will further dive into this point when we discuss the four-step training method. The second point—'B'—says that key points define those small pieces. These pieces may correspond or not to the major points as defined on Wednesday.[†] It is the trainer's responsibility to decide. The only question at this stage should be, 'Is this going to make the training easier?' Obviously, most of time there will be correspondence unless there is a need to dive into a specifically complex task. This is exactly the case we will be treating when we do T-shirt folding training. So far the T-shirt folding has been considered a major step. To ease the training we will have to split this operation in even smaller steps—five exactly—that will be taught piece by piece. Those key points might be similar to, or different from the ones included in the Operator Work Instruction (Figure 4.9). You only include key points that you would like to teach workers. As defined previously, key points are any action that is instrumental to the success of a task, whatever the field: safety, quality, productivity, or cost. However,

---

[*] Chart extracted from Improving Standardized Work: Writing Standardized Work Forms.
[†] For more details see Improving Standardized Work: Writing Standardized Work Forms.

32 • Implementing Standardized Work

Home Appliance Inc.	Operator Work Instruction		Version 00	Plant:		Team 1		Document No.: 1
Reference:	Title:		Area	T-shirt packing		Mch		Rev. Number 1
								Page: 1/1
No.	Major Steps	✚ = Safety	◆ = Quality	● = Tip	Temps:	Drawings, Photos, etc.		
1	Pick up and inspect T-shirt	✚ ◆ ◆	– Sleeve and garment hems must be flat and wide enough to prevent curling – The neckline should rest flat against the body – The neckline should recover properly after being slightly stretched			[1] Neckband — Bond on sleeve hem — Garment hem		
2	Load, start, and unload press	✚	Maintain your hands out of the press while it is closing			[2] Start button — Keep your hands in the hachured area		
3	Fold and inspect T-shirt	◆ ◆	Inspect the overall quality of the folding Check that the logo is visible on the front side of the T-shirt			[3] Logo		
4	Store T-shirt							
Author	Signature/Date	Verification		Approval		Signature/Date	Signature/Date	Comment
Name:		Quality: HSE: BIE:	Signatures/Dates	Name: Group Leader		Operator		
Function:						Function:		

FIGURE 4.9
An Operator Work Instruction sample.

unlike the Operator Work Instruction (Figure 4.9), the JBS must include an explanation for the presence of this key point (Figure 4.8). The reason is quite straightforward: people tend to adhere more to instructions when they understand the reason. This is what is written in point 'C.' Please note that the explanation column constitutes a big difference between the JBS and the Operator Work Instruction. Point 'D' specifies that there is no standardized format. Everyone can draw the format that best fits her or his preparation and teaching style. For example, as specified in point 'E,' a key point can be illustrated by drawings, pictures, or even audiovisuals if the JBS is digital media." Daniel seized the opportunity to discuss the use of technology in Lean practice. "Please remember that you should not restrain from using technology when it makes sense. Using audio-visual material is okay when it is not an end in itself. Technology should be a means to magnify your process, not the other way around. There is a famous quote from Bill Gates[*] that summarizes quite well this point, 'The first rule of any technology used in a business is that automation applied to an efficient operation will magnify the efficiency. The second is that automation applied to an inefficient operation will magnify the inefficiency.' I guess some of you have heard that, but it is always worth reminding."

People tend to adhere more to instructions when they understand the reason.

According to experts, the best way to teach people to perform an operation is to break it down into small chunks first, then teach it piece by piece, before putting the whole operation together.

Daniel stopped, dropped an eye on the room, then refocused on the subject. "Now before we discuss training methodology, I would like to introduce to you a very old document (Figure 4.10). It's a TWI card underlining key points about training. There are lot of books and methodologies out there on training, but we will focus on the TWI way. Although it's a very old document dating from WWII, I have found it to be very simple and effective. Also, the goal of this module is not to train each of you to become an education expert. I simply want to give you

---

[*] Former President and CEO—Microsoft.

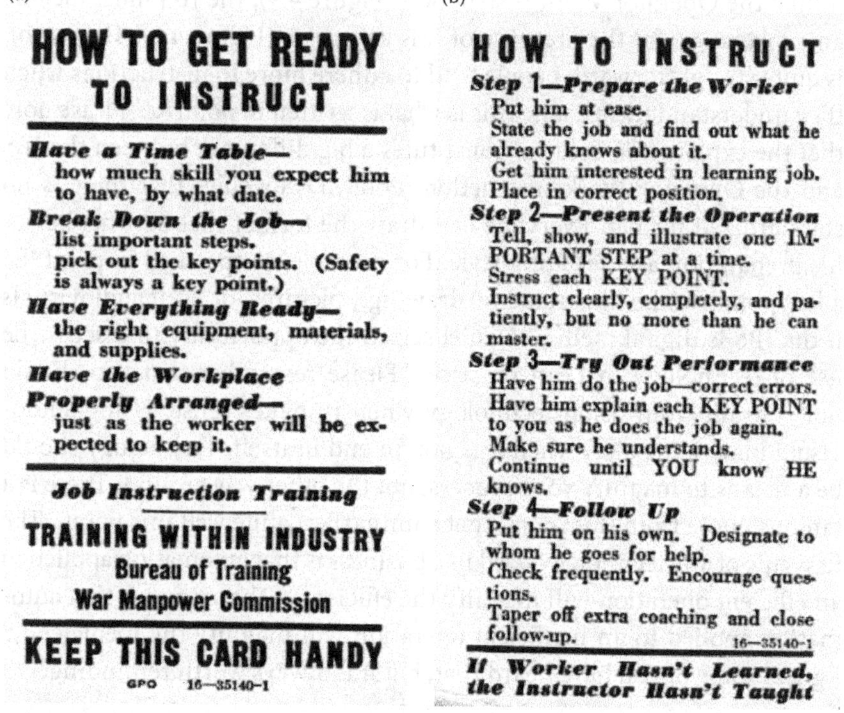

34 • *Implementing Standardized Work*

(a)

HOW TO GET READY TO INSTRUCT

Have a Time Table—
how much skill you expect him to have, by what date.
Break Down the Job—
list important steps.
pick out the key points. (Safety is always a key point.)
Have Everything Ready—
the right equipment, materials, and supplies.
Have the Workplace Properly Arranged—
just as the worker will be expected to keep it.

Job Instruction Training

TRAINING WITHIN INDUSTRY
Bureau of Training
War Manpower Commission

KEEP THIS CARD HANDY
GPO   16—35140-1

(b)

HOW TO INSTRUCT

Step 1—Prepare the Worker
Put him at ease.
State the job and find out what he already knows about it.
Get him interested in learning job.
Place in correct position.
Step 2—Present the Operation
Tell, show, and illustrate one IMPORTANT STEP at a time.
Stress each KEY POINT.
Instruct clearly, completely, and patiently, but no more than he can master.
Step 3—Try Out Performance
Have him do the job—correct errors.
Have him explain each KEY POINT to you as he does the job again.
Make sure he understands.
Continue until YOU know HE knows.
Step 4—Follow Up
Put him on his own. Designate to whom he goes for help.
Check frequently. Encourage questions.
Taper off extra coaching and close follow-up.
16—35140-1

If Worker Hasn't Learned, the Instructor Hasn't Taught

FIGURE 4.10
The two faces of a TWI (Training within Industry) card summarizing the main points to follow for an effective training: preparation (a) and instruction (b).

some quick tips to help you become a good enough trainer to implement Standardized Work. Now let us come back to the TWI card. As you can see, there are two parts, one on each face of the card: the first face (part a) of the document is about preparation for the training (Figure 4.10). This is one of the points that made you underperform when you tried to train each other. As with any task, and I would even say more than most tasks, training should be prepared to be efficient and effective. There is a famous quote from Confucius that says it very well. 'Success depends upon previous preparation, and without such preparation there is sure to be failure.' Preparation, preparation, preparation is key for success here. As you look at this document, you can see that four points are mentioned. I suggest we review them one by one."

Daniel explained that the first point: "'Have a timetable' requests that an objective be set with a related date. This is true for any project or initiative in general. The real question is, 'Where do you want to bring the trainee in terms of knowledge of the work?' Then, based on this objective, you will need to schedule your training activities. This will also help you plan for the availability of workers. As we have seen before, an operator who is being trained is not producing. Therefore you must, at least, plan for his or her replacement or avoid to include his or her contribution in your production. This is what the timetable is about. The absence of a timetable may jeopardize your training plan and feed resentment toward training activities. The next point is simply the writing of the JBS. We have already discussed this document. We will move to practice very shortly. The fourth point in the preparation of the training is the workplace. It is important that the workplace be set in the exact conditions of real operation. As we learned yesterday, work conditions are a key in performing the work properly. Again, the workplace, or very precisely, workstation layout, is part of the definition of the work. Remember what we learned from the standardized work forms: Standardized Work Combination Table, Standardized Work Chart, and Operator Work Instruction. They are all related to the workstation layout. So if you give training in a workplace that is different from the real workplace, you are training the worker to a different job. Be mindful, it is more important than most people think. I will come back to this point with a specific, frequently committed mistake (or 'FCM,' as I call it)."

Remember what we learned from the standardized work forms: Standardized Work Combination Table, Standardized Work Chart, and Operator Work Instruction. They are all related to the workstation layout. So if you give training in a workplace that is different from the real workplace you are training the worker to a different job.

Key take-aways
- Preparation is key in the training
- People tend to adhere to key points when they know the reason
- The most effective training is achieved by breaking down the operation to be taught into small chunks first, then teaching piece by piece, before putting the whole operation together
- The key delivery of the preparation phase is the JBS
- Reminder: The JBS is a free-format document, wherein tasks are broken into small pieces characterized by key points (safety, quality, cost, and productivity), an explanation for each key point and a picture/drawing or audiovisual that illustrates key points

FIGURE 4.11
Key take-aways from the preparation for training.

Daniel asked the room if there were any questions. Since nobody intervened, he asked if anyone wanted to summarize the learning before they dove into the four-step training method. Steve, the industrial engineering manager, offered to summarize. He moved to the closest paperboard and wrote down a summary in a few bullet points (Figure 4.11).

5

The Four-Step Method

Daniel started, "The four-step method is called such because the training process includes four steps (Figure 4.10). Let us go through each step. The first is about preparing the trainee. First and foremost, create the conditions so as not to be disturbed by any interruptions whatsoever: telephone, mail, SMS, people ... Keep a natural tone. Greet the trainee. Introduce yourself. At this point you need to get him or her interested in the training. The best way to do this is to show interest in the person. It means putting him or her at ease, then presenting the job (describe the purpose and the procedure) to see what the person already knows. In a sense, whenever possible you will try to link the instruction to what the trainee already knows. Specify training steps and expected duration. Invite the trainee to ask any questions or make comments. The final point of this step consists of placing the trainee in the correct position for the training. As discussed previously, this means being in the real working conditions of the task for which he or she is being trained. The one advice that I will give you here is to avoid placing the trainee in front of you. Remember, I told you that I will discuss an FCM.[*] This is it. This position adds extra effort as the trainee needs to mirror everything you do (Figure 5.1). The easiest position is side-by-side as you can see on this chart (Figure 5.2). This is actually a simple and common mistake that has big consequences."

"The second step of the four-step method consists of presenting the job to the trainee chunk by chunk" (Figure 4.8), Daniel asserted. "The main point here is to avoid overloading the trainee. I think this is pretty clear to everyone here. The best way to achieve this is to do it key point by key point. Each word you will find on the TWI document is important (Figure 4.10). Instruction should be clear, complete, and given patiently.

---

[*] Frequently committed mistake.

38 • *Implementing Standardized Work*

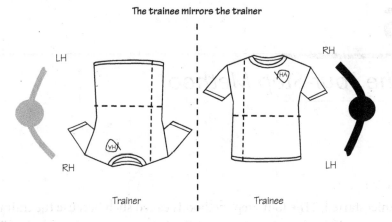

FIGURE 5.1
Face-to-face positioning makes the training harder because it imposes extra effort on the trainee.

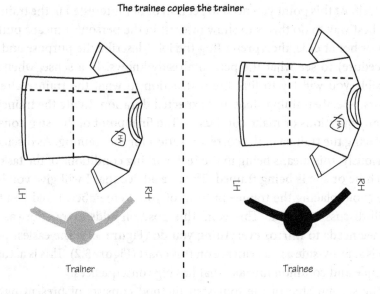

FIGURE 5.2
Side by side positioning makes it easier for the trainee to copy.

Now, how do you make instruction clear and complete? Well, by taking the necessary time to prepare and work on the key points included in the Job Breakdown Sheet. So those two previous TWI requirements depend a lot on the preparation of the trainer. The third step is about patience. It makes complete sense when you remember that TWI's motto is 'If the worker hasn't learned, the instructor hasn't taught.' How do you understand this sentence?" John, the engineering manager, raised his hand and explained that the responsibility is laid on the instructor to make sure that the trainee has understood. Therefore, it is in the instructor's interest to be patient enough to convey the knowledge and ensure the worker's performance. Daniel thanked John and dove into the third step of the TWI's four-step method (Figure 4.10). "Now comes the time to let the trainee do the job. You need to be careful that he or she is doing exactly what he or she is supposed to do. Please correct errors when they happen and make sure that he or she explains what he or she is doing every time he or she does it. There is a mix of tutoring and retrieval practice, which consists of asking the student questions several times about what he or she has learned. In effect, every time you ask the trainee to explain what he or she is doing, and every time he or she does it, you are pulling information from his memory. According to experts[*] this is a powerful way to make it stick in the memory. I will come back to this point later when we discuss scientific results justifying the effectiveness of the four-step method."

> TWI's motto is, "If the worker hasn't learned, the instructor hasn't taught."

"Finally, the fourth and the last step of the four-step method is mainly about follow-up. Indeed, you first need to thank the trainee for his dedication to the training. The basic idea in this step, which is also pulled from the power of tutoring, is to make sure the person who received the initial training is not left alone without any person to refer to in case of a problem. Remember, as we discussed previously, the worker is not at 100% of his performance after the training session (Figure 4.6). He or she needs more practice at the workstation. By perfectly balancing your input, your mission here is to lead the trainee, step by step, to full or quasi-autonomy where you will only need to coach him or her." Steve, the industrial engineering manager, raised his hand and commented that there was an evolving process quite close to what Daniel taught them regarding situational

---

[*] J. Karpicke and J. Blunt. Retrieval practice produces more learning than elaborative studying with concept mapping. *Science*, 331, 772, 2011.

leadership.* "I see the idea of the four types or steps of leadership in the process that are used to bring the learner to complete autonomy: directing, supporting, coaching, and delegating." Daniel agreed and congratulated Steve for his insightful comments and moved to the next point.

"TWI is a powerful tool." Daniel stated, "There are reasons for that. I guess things that were included in the methodology that make it so powerful were done empirically. For a long time, here is what the situation has been: everyone knew that it worked, but apart from the proof-by-result no one was quite able to scientifically explain why. The good news is that recent research works have been able to explain, in a rigorous way, some of the key factors of the success of TWI. This is what I would like to share with you now. Remember what I said, 'People adhere more when they understand why.' This is actually what I am trying to apply here on the TWI training. Well, as you know by now, one of the key principles of the TWI is to break the instruction into small chunks defined by key points that will be instructed one at a time before moving to the next stage." Daniel started, "You should not give more than the trainee can master at one time. This idea is called mastery learning. It essentially requests the learner to fully master a given amount of knowledge before moving onto a more advanced topic. As you can see on the chart there, this practice gives a better result than classical training (Figure 5.3). This result is based on an approach developed by educational psychologist Benjamin Bloom.† How to read this chart? Well, suppose that in a classical class, represented here by the bold continuous black bell curve, a student fails when her or his score is on the first half of the curve. The rate of an average student is 50%. When mastery learning is applied, the overall distribution of scores moves to the left as is shown on the gray curve. This means that the average student increases to 84%. Now the interesting point here is what comes when tutoring is applied, as we saw previously in the fourth step of the four-step method. In this case, as you see on the dotted curve, the average student's rate jumps to around 98%. The interesting point here is that as the average rate increases, the gap between the best and the worst student shrinks drastically. This certainly reminds you what we did on the first 2 days. The rule

---

* Situational Leadership is a leadership theory developed by Paul Hersey and Ken Blanchard. More details can be obtained in P. Hersey, K.H. Blanchard, and D.E. Johnson, *Management of Organizational Behavior: Leading Human Resources*, Prentice-Hall, Upper Saddle River, NJ, 2007.
† B. Bloom. The 2 Sigma Problem: The search for Methods of Group Instruction as effective as One-to-One Tutoring. *Educational Researcher*, 1984.

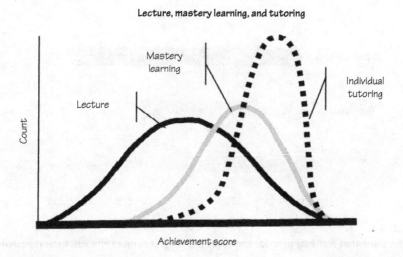

FIGURE 5.3
The 2 Sigma problem: Individual tutoring increases the pass rate of an average student from 50% to 98%.

stays the same and is coherent with our empirical observation: 'an effective training not only increases the rate of the average operator, it also reduces the dispersion within the whole population.'" Daniel paused for a few seconds and proceeded. "I hope you still remember what I told you at the beginning of this training[*]: dispersion reduction often goes with improvement of the average rate."

> *An effective training not only increases the rate of the average operator, it also reduces the dispersion within the whole population.*

Daniel moved to the next chart on retrieval practice. "There are other characteristics that make the four-step method very efficient. As you can see on this chart,[†] retrieval practice shows better results (Figure 5.4). This is the result of rigorous study. We have already discussed this point. I essentially say that the more you try to pull knowledge from your mind, the more you retain it. I will not persist. I just wanted to share with you

---

[*] For more details refer to the first book of the series: *Implementing Standardized Work: Measuring Operators' Performance.*
[†] J. Karpicke and J. Blunt. Retrieval practice produces more learning than elaborative studying with concept mapping. *Science*, 2011.

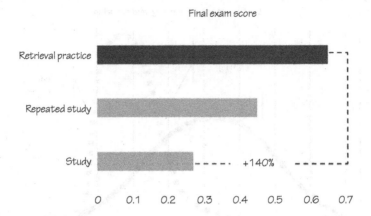

FIGURE 5.4
Studies show that retrieval practice improves learning. It is more efficient than repeated study.

a chart plotted by scientists* that proves the extent of retrieval practice's effectiveness.

The next chart is on the power of active learning; it shows that when the instruction is active, the attendance or the interest is up, engagement is improved, and as a result, learning is substantially improved" (Figure 5.5). Thomas raised his hand and commented. "This chart confirms what we already saw previously. I mean the chart over there where we can see

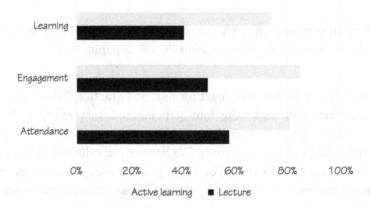

FIGURE 5.5
Studies show that active learning increases attendance, engagement, and learning.

---

* J. Karpicke and J. Blunt. Retrieval practice produces more learning than elaborative studying with concept mapping. *Science*, 2011.

that participatory methods are the ones with the highest level of retention rates" (Figure 4.6). Daniel acquiesced, thanked Thomas, and, commenting on the TWI card, he pursued. "It's really amazing! Think about that, this is a document dating from World War II that already included a few things that today's learning experts consider critical for a successful training."

When the instruction is active, the attendance or the interest is up, engagement is improved, and as a result, learning is substantially improved

Before concluding on the four-step method, to illustrate its power, Daniel wanted to share a chart with his colleagues. He explained that the chart was a practical version of Figure 3.2. It was based on data collected in one of the company's Polish plants. They were able to plot on the same chart what he called an unstructured training and a structured training (Figure 5.6). He explained, "We called unstructured training a situation where the worker was trained by a more experienced colleague who actually knew the job very well, but did not use methodology in his training—black bars. In the structured training case, the trainer used an approach similar to the four-step method—gray bars. Also, in the case of unstructured training, the worker was rushed to the line, while in the case of structured training,

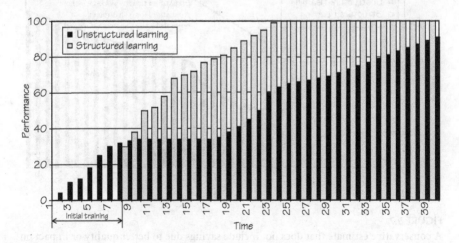

FIGURE 5.6
This chart, based on data collected from the shop floor, shows that structured learning helps to save money due to better performance and fewer errors (reduced rework time).

he or she received the necessary initial training off line. The second worker started producing 8 TU (time units) after the first one (Figure 5.6). As you can see, within the observation period, there is a big gap between the two workers' performances, which translated into big savings when the operator was trained in a structured way. We made a rough estimate of the gap that you can see on this chart (Figure 5.7). We obtained a solid 38% savings when the worker was trained properly versus when he received casual training. Please note that 38% is an estimate on the period of observation. One should expect the gain to be greater ahead. Chances are very high that there will always be a gap between the two operators as in most of cases the ill-trained operator never reaches 100% performance. Therefore, saving will increase over time. Anyway, let us suppose conservatively that we are saving 38%, which is not a small amount. Think about this; contrary to what most people think, training is not a waste of time and money, but a way to actually save money." At this point, Daniel stopped and glanced at the room. Sarah raised her hand and commented on the charts. She mentioned that they were the first hard proof she had ever seen clearly showing the benefit of training. The time was right to take a break. Before leaving, Thomas made a quick summary. In the summary he underscored the power of the four-step method. He called it an old practice that has been powerful in training people in industry. He reminded participants of

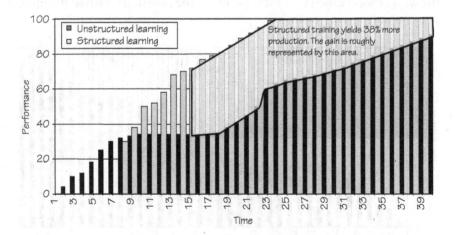

FIGURE 5.7
A conservative estimate that does not include savings due to better quality or impact on other workers shows a solid 38% savings, represented by the gray area. Also the estimate corresponds to the period of observation. The saving is expected to increase over time as in most of cases the ill-trained operator never reaches 100%.

the four steps of the method: (1) prepare the worker, (2) present the operation, (3) try out the performance, and (4) follow up. He reiterated some of the key points that make the method very efficient. As he explained, it was based on four powerful aspects: the practice or active learning, chunk-by-chunk instruction, retrieval practice, and tutoring. He commented on the retrieval practice as follows. "The basic idea here is that the more you try to 'retrieve' what you have learned, the more you remember. I would not imagine that such a simple action could yield so much." As a final point, he mentioned the last chart, which showed a shop floor example of the benefit of a structured training like the four-step method. "I could not anticipate such a big impact: 38% is quite impressive," he concluded.

Daniel thanked Thomas and signaled the beginning of the break. As the week ended, participants were getting exhausted from a long intensive week of work. Therefore, Daniel decided to give more breaks than he would do normally. This one was the second break of the morning and probably not the last. Before they left, he told the group that when they returned they would be given a concrete example based on a new T-shirt folding method. They would have to write the Job Breakdown Sheet, prepare the training, and then train each other using the four-step method.

> *We obtained a solid 38% savings when the worker was trained properly versus when he received casual training.*

6

Practicing Training: The Second Chance to Get It Right

When the participants returned, Daniel showed them a new method of folding T-shirts, and asked each group to prepare corresponding training and to come up with a Job Breakdown Sheet. The method, he explained, could be seen on the internet on several YouTube videos. This would help reduce the folding time drastically, to 2 or 3 s. Though, he insisted, "As we saw yesterday* you do not need that folding method to reach a cycle time below the takt time, which was our primary process improvement objective. Yesterday we managed to reach 17 s, 3 s below the takt time of 20 s, without this method. More generally, once your time to fold, inspect, and store is below 10 s of the press machine time, if you are using two T-shirts,† the folding is no more on the critical path of the cycle time (Figure 6.1). Based on the Standardized Work Combination Table of the improved process we discussed yesterday (Figure 6.1), whatever your folding time, the cycle time will not go below 14 s unless you improve major steps 1 and 2. Some might wonder why I am lingering so much on this point. Well this is an opportunity for me to caution you against the 'improvement-for-improvement's-sake' reflex that I see so often in plants. Now let us focus on the tasks to be done." Daniel then distributed a blank Job Breakdown Sheet. He gave an additional explanation on each column of the document. "The first column is for major steps what we learnt for the Operator Work Instruction applies here as well. It should be expressed as a verb and an

---

* More details are available in the third book of the series: *Implementing Standardized Work: Process Improvement*.
† In the improved method presented in the aforementioned third book, the Standard Work In Process (SWIP) contains two T-shirts. The operator folds the first T-shirt while the second T-shirt is being folded. For more details see *Implementing Standardized Work: Process Improvement*.

STANDARDIZED WORK COMBINATION TABLE

TITLE	T-shirt Packing – Improved –			BY: Team 1		TAKT TIME:	20 seconds	COMMENTS
						Volume:	17,280 T-shirts per day	2 t-shirts in the work in process
						DATE:	June, 2012	

N°	Major steps	TIME			TIME GRAPH (seconds)
		MANU	AUTO	WAIT	WALK
1	Unload press (t-shirt 1)	1	0	0	0
2	Pick up & inspect new t-shirt, load and start mach. (t-shirt 2)	3	10	0	0
3	Fold, inspect and store ironed t-shirt (t-shirt 1)	13	0	0	0
	TOTAL	17	10	0	0

MANUAL ▬▬ AUTO ---- WALKING ∿∿∿ WAIT: ☐

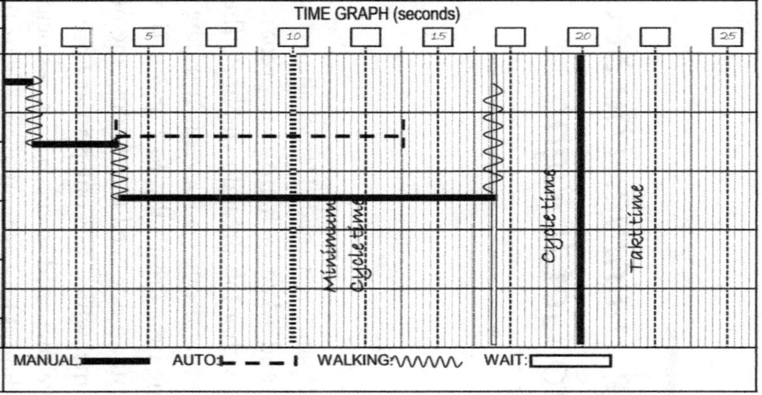

FIGURE 6.1
Standardized Work combination table of the improved operation.

object, not more. Keep it simple. The second column includes key points. The previous explanation applies as well. Remember we are talking about anything that can (1) make or break the job, (2) injure the worker, or (3) make the work easier. In the thin adjacent column there is room for symbols, please refer to the bottom of the document to pick the one that corresponds to the key point: safety, quality, making the job easier, or critical quantity check. The next column is about the 'reasons.' Please describe your reasons in a very simple way as well. Use short sentences, ideally a verb plus objects. The last column, as you know by now, is reserved for illustration. This could be hand sketches, diagrams, parts, layouts, or digital picture if available. That's all from my side.... Is everything clear to you? If so, let's get started." Daniel handed out to each group a memory stick including a video of the new folding method that they would be able to play as often as necessary to capture the new method and build their Job Breakdown Sheet. Daniel got several questions while walking around.

Key points are about things that can (1) make or break the job, (2) injure the worker, or (3) make the work easier.

After 20 min, Daniel called the group in order to share their solutions and obtain their feedback. As always, he convened all participants around the group that produced the best result and asked the best group to present their work and explain to others. This approach was one of Daniel's tips for an efficient training. Whenever it was possible, he always tried to use some participants to train the rest of the group. The group that produced the best response was Eric's. Therefore, Daniel asked the young Excellence System manager to explain his group's result to the rest of the participants (Figure 6.2). Their JBS included five major steps. Each step had one to three key points, and each key point had a justification. In the last column of the document they illustrated some key points with clear drawings that showed very well what the trainee was supposed to do. Daniel took the opportunity to underscore the importance of the drawing in the last column. "Guys, you must make clear drawings. I do understand that not everyone is good at drawing. If this is your case, please use pictures. If you decide to use pictures, make sure that they are clear and unambiguously illustrate the task. Now let me say it again, there is no problem in using a video if you can afford digital material. It can be a very powerful training medium. You all know this famous quote: 'A picture is worth a thousand words', right?

50 • Implementing Standardized Work

Job Breakdown Sheet

Description of the task: **T-shirt folding**
Parts (UPN, describe the parts): /
Tools & supplies required: /
Safety equipment required: /

List Common Key Points:

	Major Steps	Key Points		Reasons
WHAT?	A logical segment of the operation that advances the work.	HOW?	Things in important steps that will: 1. make or break the job 2. injure the worker 3. make the work easier	WHY? List the reasons for the key points
1	Lay T-shirt flat	a - Front up b - Whole T-shirt (TS) completely flat		a - Avoid folding inside out b - Avoid twists and wrinkles
2	Pinch T-shirt (Two-hand)	a - Draw imaginary line parallel to TS border b - Left hand (LH) pinches middle of line c - Right hand (RH) pinches top/shoulder		a - Folded TS won't be trapezoidal. b - Upper, lower parts fold equal length c - Ensure proper folding
3	Grab T-shirt (Right hand)	a - Right hand turns around left hand b - Right hand grabs bottom / hem		a - Flops sleeve and upper band into TS b - Achieve proper folding
4	Fold T-shirt, part 1	a - Pick up and lift T-shirt b - Uncross arms		a - Make fold b - Bring front outside / toward operator
5	Fold T-shirt, part 2	a - Lay sleeve on the table b - Fold sleeves below, check distance to collar		a - Make folding possible b - Fold symmetrically

Key Point reminders: ● Critical check or Inspection ◆ Could cause injury △ Quantity Check ● Makes the job easier

Training Aid: (put hand sketches, diagrams, parts, or layouts here. Insert digital picture if available.)

Owner of this document: _____ Page: __ of __ Rev: __ Date: __

FIGURE 6.2
Job Breakdown Sheet of the best group. Key points are related to each major step. Each key point is explained and illustrated by drawings.

Well, I have built my own version that says: 'A picture is worth a thousand words, a video is worth a thousand pictures, and practice is worth a thousand videos.' I think it is pretty clear; no need to explain, right?" Everyone acquiesced. Daniel then carried on, "Digital material—like video—is good stuff!" However be careful not to get absorbed by the time or energy needed to build such material. Also, it might reduce your ability to make changes to update the results of continuous improvement workshops, which in practice should be the rule in our plant. You need to reach the right balance. However, it is always preferable to have a dirty and not so sexy document that conveys that latest best practice than a nice video that describes a dated method. Whenever possible, keep it simple and effective."

> *A picture is worth a thousand words, a video is worth a thousand pictures, and practice is worth a thousand videos.*

> *It is always preferable to have a dirty and not-so-sexy document that conveys the latest best practice rather than a nice video that describes a dated method. Whenever possible, keep it simple and effective.*

There were several questions to which Eric and his group responded quite well. Then Daniel requested that each group align itself with Eric's group result and then move to the next step, which consisted of using the Job Breakdown Sheet to train. He showed a chart to illustrate the best way to perform training (Figure 6.3). "Let's say you need to train someone for a task that you decided to split into four chunks. You would need four sequences. In the first sequence, after the trainer explains and demonstrates the first piece, the trainee performs Piece 1, while explaining what she or he is doing. The trainer gives feedback followed by corrective procedures. The trainee repeats as often as needed until the work is mastered. Thereafter the trainer completes Pieces 2, 3, and 4. In the second sequence, after the trainer explains and demonstrates the second Piece, the trainee performs Pieces 1 and 2 again, while explaining. The trainer gives feedback followed by corrective procedures. The trainee repeats as often as needed until the work is mastered. Thereafter the trainer completes Pieces 3 and 4. In the third sequence, after the trainer explains and demonstrates the third Piece, the trainee performs Pieces 1, 2, and 3 while explaining. The trainer gives feedback followed by corrective procedures. The trainee

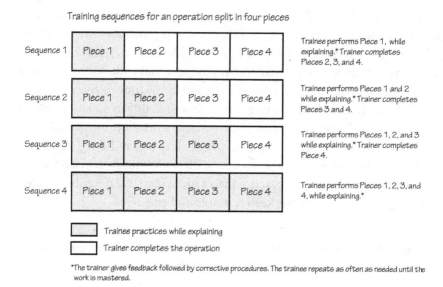

FIGURE 6.3
Illustration of piece-by-piece training activities.

repeats as often as needed until the work is mastered. Thereafter the trainee completes Piece 4. In the fourth sequence after the trainer explains and demonstrates the fourth Piece, the trainee performs Pieces 1, 2, 3, and 4, while explaining. The trainer gives feedback followed by corrective procedures. The trainee repeats as often as needed until the work it is mastered. The trainer has nothing to do but to congratulate the trainee. As you may see, the more pieces you have, the more sequences it will take to train the person. But in the end, this will guarantee the best result. So avoid taking shortcuts. If not, you risk skimping the work.".

The practice went on. The most difficult point to convey was the step-by-step training and making sure that the trainee had mastered a step before moving to the next one. On the trainee side, the challenge was to say what they were doing while doing so. However, this time the results were different, and the participants seemed to enjoy the results they achieved.

As Daniel was about to conclude and move to auditing, a question popped-up about the need, and the tool to be used, for evaluating the training. Daniel responded, "Well, let me provide you with a few elements as a response. First of all, you ought to remember that this is not a training about training. My goal here, as I stated previously, is to provide you with some quick tools and knowledge to help you train an operator to Standardized Work. The second point is that our focus here is not evaluation, as we consider it your responsibility to do whatever it takes to teach the worker. Remember the TWI motto: 'If the worker hasn't learned, the instructor hasn't taught.' The third point is that we believe, as they do at Toyota, that 'the right process will lead to the best results.' Therefore to be the most effective, we prefer to spend the least time dedicated to the training on its evaluation to make sure you master the right training process itself. The fourth and last point; I can actually provide you with an easy and quick tool to make evaluations." Daniel went to the board and started drawing while explaining (Figure 6.4). "I suggest you rate the outcome of the training on three levels that could be green, orange, and red corresponding respectively to success, mixed result, and failure. The circles on the left show the estimated level of worker performance. As explained previously, the highest level you might expect at this level is 75%. In this case, the worker is ready for the workstation. The next possible situation is when the training is not completely successful and you consider that the worker has only mastered 50% of the total work, you may rate the training as 'orange.' There are two options at this point: (a) retraining or (b) continue at the workstation with very close team leader supervision. The last outcome, which is failure, means that the worker is not ready for the workstation. In this case you should primarily check the training conditions and the trainer capability before jumping to worker behavior. Again, here as always, the moto is 'If the worker hasn't learned, the instructor

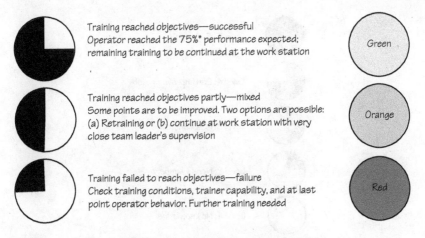

*75% is the highest level of performance that should be expected from training session

FIGURE 6.4
Simple evaluation grid to be used at the end of the training.

hasn't taught.' There will be a need for another training unless it is diagnosed that worker behavior is the root cause of failure. The assumption here, though, is that the worker meets the necessary prerequisites to take the training."

Before wrapping up this part of the training, Daniel wanted to add a few caveats. "Please let me underscore that the evaluation grid we just discussed (Figure 6.4) is a simplified version. In most companies, you will find a grid with 5 levels, which are from the lowest to the highest: 'No knowledge,' 'Needs support,' 'Is autonomous,' 'Can improve the job,' and 'Can teach' (Figure 6.5). Also, I would like to add that if you are training a worker on a set of tasks, the usage is to provide 3 columns for evaluation: a column that specifies the target, a column for self-evaluation, and a third one for the instructor's evaluation. This is an example from a competitor's plant (Figure 6.6). We need the trainee to evaluate himself or herself to make sure that both parties have a say. A gap between self-evaluation and the instructor's evaluation should not be considered a problem; rather it is an opportunity for the instructor and the trainee to discuss and share their perceptions."

Daniel summarized the key take-away in one point: the training aid or illustration, "You have the right to be creative and also to use technology. However, think about balancing this with the ease of updating your document quickly. You need to find the best equilibrium point here." He also

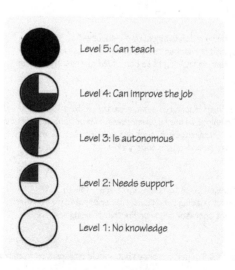

FIGURE 6.5
In most companies the evaluation grid has 5 levels.

| | | | | | No knowledge ○ | Needs support ◐ | Is autonomous ◑ | Can improve ◕ | Can teach ● |

Photo | Name
Qualification

No	Subject	Educational tool	Trainer	Time	Evaluation target	Self-evaluation	Instructor's evaluation
1					⊕	⊕	⊕
2					⊕	⊕	⊕
3					⊕	⊕	⊕
4					⊕	⊕	⊕
5					⊕	⊕	⊕
6					⊕	⊕	⊕
7					⊕	⊕	⊕

	Comments	Signature
Trainer		
Group leader		
Manager		

FIGURE 6.6
Example of an evaluation sheet that includes the target, self-evaluation, and the instructor's evaluation.

added, "It is really important to train chunk by chunk, and while doing so you need to make sure that the trainee is always saying what he or she is doing; it goes back to the power of retrieval practice we studied previously" (Figure 5.4).

Daniel concluded the part on training. It was lunch time. The group took a break for an hour-long lunch and was soon back to the room for the second subject of the day: auditing.

7

Introducing Auditing

As always, Daniel used lunchtime to get some feedback about the reception of the training. Over the week, he would make sure that he had the opportunity to sit at a table with different people to be able to capture what he called the heartbeat of everyone. Years of teaching and leading workshops had taught him how important it was to detect weak signals from all participants. He had also noticed that most of the time, people who speak the least are the best observers, who tend to provide the sharpest assessment. Therefore, Daniel would generally use breaks to initiate some chat with the people he called the "quiet brains."

Daniel also took some time to check in with Thomas regarding his progress with the unions. Early reactions toward the competitiveness plan were mixed. "I think smart people are starting to understand that there is a crisis out there. The situation is changing very quickly. We cannot waste our energy fighting each other while we should be assembling our forces to become more competitive and save the plant. I do understand there is some distrust based on history. My first mission is to rebuild this trust." Daniel then wanted to know more about Thomas's approach to rebuild the trust. "Well Daniel, since I took over this position, I have been trying classical techniques, which consist of getting people away from more conventional places. For example, I asked Sarah to launch a contest where each of our associate's kids would produce a drawing based on the theme of safety in our plant. The participation rate was very high, and the jury, an external panel, selected and ranked 10 drawings. We then invited all of the families to attend a giant barbecue in our garden here to celebrate the winners. This was an opportunity to get our peoples' heads out of the river of gloom, build a sense of 'we are all in this together,' and ultimately foster some trust among each other. Indeed, do not get me wrong; years of strained relations with unions will not be simply erased with a big barbecue party. I know that it takes more than that. It is simply a small stone to start rebuilding trust. It is really important that we start working hand in hand with unions, and through them, all employees. Everyone must feel secure. I had a discussion with the regional VP of manufacturing and asked him to give me the permission to announce that whatever happens to the plant, nobody will be fired for economic reasons in the next 5 years. He was a little bothered since he wanted to keep his options open. I think after years of bad results from this plant, he has lost his patience. I therefore had to put my resignation on the balance. 'My conviction,' I told him, 'was that it is important to fill the trust gap between the plant management and employees.' Think about the disarray this plant has been in for several years. We are asking people to make important sacrifices. In return, they should have some assurance that it will pay off."

Back in the training room, Daniel started by making the point he made before about auditing being the "C" and the "A" of PDCA, which is the backbone of Standardized Work deployment. He then seemed to shift completely to another subject when he asked Steve, the IE Manager, if he would like to draw the organizational chart on a paperboard. Steve, who knew this by heart, accepted, went to a paperboard and quickly represented the plant organization chart (Figure 7.1). To save time, Daniel asked him to only represent the part related to production hierarchy. He started with the Plant director. Below the Plant director there were Supervisors, and below the Supervisors, there were Group leaders, and below Group leaders there were Team leaders.

Like most companies, the organization chart here was a casual mix of Toyota's and the company's own job position glossary. Unlike in Toyota's organization, the Team leader had a real hierarchical and managerial role. He or she was a sort of first line manager. In effect, this was the appellation the company used before its conversion to Lean. The first line manager was in charge of a team of 30 operators on average. Then position names were changed, and the team size was reduced to get closer to Toyota's organizational structure. The new target of seven operators for a Team leader was marshaled throughout. As it happens sometimes in Lean deployment, people wanted to surpass the master.* This new organization had been routinely assaulted by "numbers guys." One of them, from the headquarters controlling team, calculated that if the company would move from its average of seven persons per team to an average of 14, it would reduce its

---

* At Toyota there is generally a Team leader for a group of 5–8 operators.

Introducing Auditing • 59

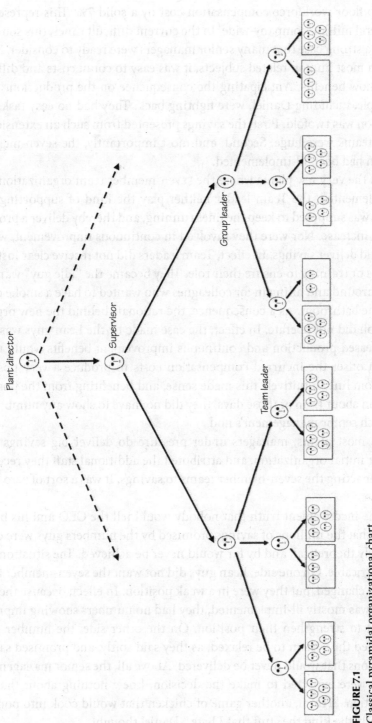

FIGURE 7.1
Classical pyramidal organizational chart.

shop floor workforce compensation cost by a solid 7%. This represented several million company-wide. In the current difficult times, this sounded like a strong case that many senior managers were ready to consider. Here, as in most human related subjects, it was easy to count costs and difficult to show benefits. Anticipating the consequence on the production, Lean people, including Daniel, were fighting back. They had no easy task. The reason was twofold. First, the savings presented from such an extension of the teams were huge. Second, and most importantly, the seven-member team had been ill-implemented.

In the very few plants where the seven-member team organization was implemented, the Team leader neither play the kind of supporting role that was supposed to keep the lines running, and thereby deliver a production increase. Nor were they involved in continuous improvement, which would deliver savings. In effect, Team leaders did not receive clear instructions or training to ensure their role. They became the "idle guy" wandering around and filling in for colleagues who wanted to have a smoke or go to the bathroom. As a consequence, the rationale behind the new organization did not operate. In effect, the case made by the Lean guys was that increased production and continuous improvement benefits would more than offset the incurred compensation costs to produce a very positive bottom line. Intuitively this made sense, and benefiting from the positive mood about Lean of those days, they did not have to show any numbers to clinch senior management's nod.

In most plants, managers under pressure do deliver big savings kept their initial organization, and attributed the additional staff they received for enacting the seven-member teams to savings. It was a sort of zero-sum game.

The inconvenient truth that nobody would tell the CEO and his board was that the millions of savings promised by the numbers guys were completely theoretical, and by far would never be achieved. The situation was inextricable. On one side, Lean guys did not want the seven-member team to be changed, but they were in a weak position. In effect, because the system was mostly ill-implemented, they had no numbers showing improvement to strengthen their position. On the other side, the number guys wanted the system to be relaxed, as they said softly, and promised saving millions that would never be delivered. Above all, the senior management, who were expected to make the decision, knew nothing about the real situation. "Hmm, another game of chicken that would cook into politics; exactly the kind the stuff that I hate," Daniel thought.

Daniel refocused on the class, pulled from his thoughts by Steve shouting "That's it." Then he commented, "Steve, I was expecting you to draw this chart. It is absolutely correct. The problem with this chart is that it depicts something like a kingdom. Let's say an ant organization. A classical organization chart, such as an ant organization, allegorically shows a king or a queen, here it is the Plant director, whom everyone else is at the service of. Well, I am caricaturing a little bit indeed. Hmm....what I mean is that your chart does not show that the value is actually created by people on the shop floor, and therefore they are the raison d'etre of production management and other support functions." Daniel moved close to a paper board and started explaining while drawing. "My preferred way of showing the organization in the plant is the reversed chart (Figure 7.2). As you may see, the key difference here versus Steve's chart is that what is most important comes first. Operators are represented in the first row, then Team leaders, Group leaders, Supervisors, and Plant director. All those guys are there to support production. So from now on I will be calling them the 'Production support structure.' It really goes beyond simply reversing a chart. This is a matter of mindset change and the organization chart is only there to materialize it. Now, you might be asking yourself 'what is the link with auditing?' Okay, the link with auditing is that auditing is an important support activity and therefore needs to be introduced by the production support structure. I should start with the Plant director, who introduces the auditing to his or her Supervisors. In their turn, Supervisors will introduce auditing to Group leaders. Group leaders will introduce auditing to Team leaders."

> *A classical organization chart, such as ant organization, allegorically shows a king or a queen, here is the Plant director, whom everyone else is at the service of.*
>
> *My preferred way of showing the organization in the plant is the reversed chart (...) It really goes beyond simply reversing a chart. This is a matter of mindset change and the organization chart is only there to materialize it.*

Eric raised his hand to ask a question about how often auditing must be performed. In principle, Daniel responded, the more auditing the better. "Ideally one must be able to have a continuous view of reality to see

62 • Implementing Standardized Work

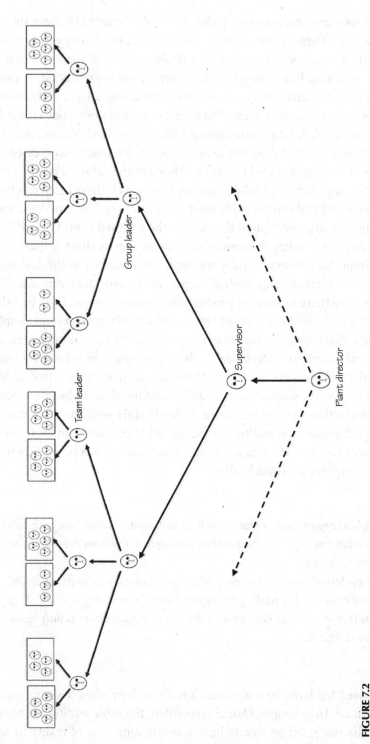

FIGURE 7.2
The reverse organizational chart embodies the idea that the management's role is to support operators in their value adding work.

any discrepancies, and act as soon as possible. The problem is the 'cost' of doing it. It takes some time, and frankly although auditing is an important activity as we agreed previously, it is just an ancillary task related to production. It is not where manufacturing people should be spending their time. There is no value adding in auditing, however auditing will help you to be highly efficient and effective in your value adding activities. Well, let me summarize; if you want frequent auditing and you are mindful of its cost in terms of time, then you need to make it easy. This means working on the procedure and the organization to make sure that you minimize the time needed for auditing. For example, auditing reporting documents should be very visual and easy to fill out. People involved, from the operator to the Plant director should play complementary roles with some overlap, not redundant ones. As you now are aware, auditing is an important step of the Standardized Work, therefore it should be introduced by the Plant director as discussed before. His responsibility and that of others in the plant's production structure does not stop there. They need to do some of their own and also 'audit auditing.'" Daniel unveiled a chart he drew while preparing the training session (Figure 7.3). "I have included in this chart how frequently I think everyone should be performing auditing and how often they should be auditing the auditing. Well, by 'auditing the auditing' I mean everyone, from Group leader to Plant director, should also make sure that the people reporting to him are actually doing their share of activity regarding auditing. At the end of day, as we shall

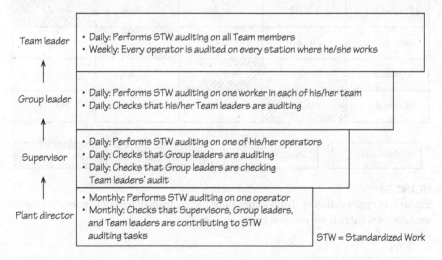

FIGURE 7.3
Production management must be involved in auditing and auditing of auditing.

see later, auditing should be entirely part of the production management activity routine. Talking about auditing frequency precisely, I would like to add that my suggestions are to be taken as a starting point (Figure 7.3). I suggest that you do not hesitate to adjust them based on the situation at hand." Daniel dove into the Team leader's role. He explained, "As you can read on the chart the Team leader needs to audit each operator of his or her team on a daily basis. However, you know that there is job rotation* in our production organization that mandates every operator move to a different workstation every 2 h. In this case, the Team leader needs to make sure that he or she has audited each operator on every job-rotation workstation. Let us take a simple example of a team of four operators who are working on four workstations. They move to a different job position every 2 h. This table shows an example of auditing organization (Figure 7.4). In this organization, on day 1, the Team leader will audit Operator 1 on Workstation 1; Operator 2 on Workstation 4, Operator 3 on Workstation 3, and Operator 4 on Workstation 2. The next day (day 2) he will audit Operator 1 on Workstation 2; Operator 2 on Workstation 1, Operator 3 on Workstation 4, Operator 4 on Workstation 3, and so on. By so doing, the Team leader makes sure that all operators of his or her team are audited in all job situations."

	Operator 1	Operator 2	Operator 3	Operator 4
Station 1	1st 2 h	2nd 2 h	3rd 2 h	4th 2 h
Station 2	2nd 2 h	3rd 2 h	4th 2 h	1st 2 h
Station 3	3rd 2 h	4th 2 h	1st 2 h	2nd 2 h
Station 4	4th 2 h	1st 2 h	2nd 2 h	3rd 2 h

| Auditing day 1 | Auditing day 2 | Auditing day 3 | Auditing day 4 |

FIGURE 7.4
Example of organization of Team leader auditing in a team of four operators working on four stations. There is job rotation every 2 h.

---

* The benefit of job rotation is mostly on the ergonomics. It allows operator to level workload and body muscles solicitation.

As you now are aware, auditing is an important step of the standardized work, therefore it should be introduced by the Plant director as discussed before. His responsibility and that of others in the plant's production structure people do not stop there. They need to do some of their own and also 'audit auditing.'

Before closing the module on auditing introduction, Daniel asked Thomas if he would like to summarize. Thomas went to the board and commented while writing. "I was really impressed by two points (Figure 7.5). The first one is the reversed organization chart. I found the idea behind it really powerful. Therefore, I will make sure that the one displayed at the entrance of the plant be changed as soon as possible. Also, in the future we will be using a reversed organization chart in all our documents. The other point is the management role, first, in introducing and then in sustaining auditing. My team and I will be doing our share." Daniel thanked Thomas and offered a break to the participants. "When you return we will discuss the auditing document."

The break was supposed to last 10 min, but it took twice the time. Daniel and Thomas had a quick discussion on the negotiations going on with the unions. Thomas had just received a phone call from the head of the union organization informing that his board had decided to support the competitiveness plan. However, to make sure that everyone has his or her say given the importance of the decision, they would maintain the poll as previously agreed. This was great news for Thomas. He had the feeling that his style and new way of engaging unions were paying off. Being of German culture, his approach to relations with the union was less confrontational.

Key take-aways

- Adopt the mindset of the reversed-organization chart:
  - Operators are the ones who add value
  - The production management and other support functions are there to support them, and make their life easy to this end
- The production support structure (production management) should
  - Carry out the introduction of auditing in the plant
  - Be routinely involved in auditing (perform some of their own)
  - Must also "audit the auditing"

FIGURE 7.5
Key take-aways from auditing introduction.

Since he joined the plant, he decided to do things the German way, not only by cooperating, but also by being more straightforward with the unions. His bet was that they love their plant; most of them have been working there for more than 20 years. Therefore, being well aware of the situation of industry in France, they would be willing to discuss with a trusted person and make tough decisions. He had always thought that their initial opposition was mostly a matter of posture and defiance. He was therefore convinced that once some trust was established, they would be able to work together. "I think they really appreciated my commitment not to fire a single person over the next 5 years. They saw this as another proof that I was serious about my embracement of togetherness." No surprise that Thomas had a big smile. He was in the mood for celebration. In effect, with the support of the unions, consulting with employees had become a formality. The anticipated outcome would be positive. Suddenly he had the impression that the hard work waged was starting to be rewarded. Daniel congratulated him, and encouraged him to keep going in that direction. Thomas, thought he owed this success to Paul, the head of unions. He explained, "You need a strong partner to build something; when you are negotiating with unions it is never good to have a weak person on the other side. It's like the tango, to be at the top you need a good partner." He then concluded, "I cannot turn around this plant alone, it's going to be a team work. If there is anything I have learnt from life, it's that little of consequence is ever accomplished alone."

8

Key Points in an Auditing Document

Daniel has been using an auditing document he adapted from an initial one found in the literature.* "Just like any such document, there are a few pieces of information that you need to fill out: who, where, when. I just handed a blank document to you (Figure 8.1) where you can find some questions you need to answer during auditing. First of all, note that the document is designed so that you only need to mark 'O' or 'X.' The mark 'O' means that the answer to the question is positive, which translates into 'no problem.' Conversely if there is any problem, then you need to mark 'X.' In this case, you have to provide more information in the comment area. The question number should precede the comment. Please note that some questions are mandatory and some others might be customized based on the specificity of the work being carried out by the operator ... Well one remark before you ask the question. Although, there is a room to be filled beside 'Operator' on the document in your hands (Figure 8.1), please keep in mind that, as any Standardized Work document, the Audit Sheet belongs to the worker position (workstation), not to the operator as an individual. Therefore it does not follow the operator as he or she moves to another workstation for job rotation." Daniel asked a few participants to read the 10 questions listed on the documents.

Question 1: "Are the Standardized Work forms up to date?" Daniel commented. "This is a housekeeping question. The idea here is that every time you use a Standardized Work form you need to check first of all that you are using the latest version. There are several ways to check; the date could be an indicator. A very old date means that the work station has not been improved for a while, which is either an indicator of lack of continuous improvement activity or an absence of updating. In both cases, it should

---

* Mike Rother and Rick Harris, *Creating Continuous Flow.* Lean Enterprise Institute, Cambridge, MA, June 2001.

FIGURE 8.1
Example of audit sheet.

be worrisome to every manager. That said, the best way to check if forms are up to date is to have a look at the workstation layout and compare it to the standardized work chart. Please note that a station can operate at a different speed or takt time. As we saw previously, the Standardized Work depends on the takt time, which means that you need to make sure that the chart you are using to carry out the auditing is the right one. I mean check that the forms correspond to the right takt time."

Question 2: "Are the Standardized Work forms visible?" Daniel commented, "Well, the rationale supporting this question is that the forms should be easy for everyone who needs to use them to do so. I need to say that more and more plants, including some from Toyota, are now considering that displaying a bunch of papers at a workstation could represent fire hazard. Some have even pushed this logic to the point where they are now keeping their forms in their computers and printing them as needed. My advice is that you need to keep your Standardized Work forms visible. The rule that applies here is 'Out of sight, out of mind.' If it is not visible, the risk is that all activities around, which are key to plant management, might simply be forgotten. Let me take this opportunity to underscore that Standardized Work forms are not placed at the workstation for the primary use of operators. Workers are supposed to be trained when they operate at their workstation. Documents are primarily there for managers who will mostly use them for auditing. What I just said is true for most tasks in industry. In some sectors like aeronautics, the cycle time of an operator can be very long. It might take more than 20 min. In this case, some operators might not be able to keep the details in their mind when they are new at the workstation. They might therefore need some visual support. Whatever the reason, if your workers need to consult a document before doing their job, then they are not working in an efficient way. You should therefore strive to find a way to break down the amount of the work before deploying Standardized Work. The justification for this approach goes back to the invention of the assembly line by Henry Ford in 1913.'"

Question 3: "Are the Standardized Work forms dated and signed by the right people?" Daniel said. "This is another housekeeping question. Every time you have the document in hand you need to have the reflex to check

---

* "On this day in 1913, Henry Ford installs the first moving assembly line for the mass production of an entire automobile. His innovation reduced the time it took to build a car from more than 12 hours to two hours and 30 minutes"—according to http://www.history.com/this-day-in-history/fords-assembly-line-starts-rolling, accessed on April 30, 2015.

if they are dated and signed by the right guys. If the document is not dated or signed, it has no value. Please remember this point."

Question 4: "Does the number of Standardized Work forms match the number of operators?" Daniel insisted, "I am sure you still remember, but let me say it again: 'one operator, one set of Standardized Work forms.' Standardized Work forms are not linked to the machine but to worker positions and several workers may work on a single machine. Do not forget that!"

Question 5: "Are the four elements of the Standardized Work represented on the forms?" Daniel added, "This question, like the previous ones, is also a housekeeping point." Daniel asked if someone in the room could recall the four elements of Standardized Work. One participant raised her hand and listed. "The Takt Time, Job Sequence, Standard Work In Process, and Key Points."

Question 6: "Is the Standard Work in Process (SWIP) correct? You basically need to check if the parts that are needed to perform Standardized Work are actually there, and placed where they should be. This is actually the very first thing to check after the previous questions, which were mostly housekeeping questions. As seen previously, if the SWIP is not respected, then the operator cannot work according to the defined standard. This is a prerequisite for Standardized Work observance. Take for example the improved T-shirt folding method. We found that there needed to be two T-shirts in the process to for the operator to be able to work according to this method and achieve the 17 s[*] we reached. Without those two T-shirts in the flow there is no way the operator could be following the improved method. Therefore the auditing stops here."

> As seen previously, if the SWIP is not respected, then the operator cannot work according to the defined standard. This is a prerequisite for Standardized Work observance.

Question 7: "Is the work being performed according to the Standardized Work forms?" Daniel specified, "This is a check of both the job sequence and key points of Standardized Work. The best form to start with is the Standardized Work chart. It may not be easy to validate the observance of key points. Therefore, at this point the checking will mostly be about job sequence."

---

[*] For more details on the improved method, refer to the third book of the series: *Implementing Standardized Work: Process Improvement*.

Question 8: "Is the worker performing at takt time?" Daniel explained, "The easiest way to check this point is to use a stopwatch to measure three cycles. If the worker is out (below or above) by an interval of 10%, then chances are high that he or she is not respecting the standard method."

Question 9: "Is the cycle time represented on Standardized Work forms correct?" Daniel explained, "This is a housekeeping question. Although the cycle time does not belong to the four elements of Standardized Work, it is important information. It defines the work conditions of the operator. More precisely, a longer cycle time can prevent observance of the Standardized Work. Remember, the Standardized Work combination table is based on the right combination of machine and worker activity. It's like choreography."

Question 10: "Is the machine running at the Standardized Work chart's cycle time?" Daniel added, "This point is simply the application of the previous point. The point here is 'is the cycle time represented on the Standardized Work forms actually implemented?.'"

Daniel stopped there and explained, "These are the main questions. Of course you can add more. For instance, you can think about adding another one regarding housekeeping on the production board: 'Is the hourly production target and actual performance written on the production board?' Again as discussed previously, if for some reason you would like to add a point that you think should be an important focus, feel free to do so. However, keep in mind that more points to check means lengthier auditing, which might end up frustrating people carrying out audits, especially if they do not see the importance. There is a real risk of distraction from vital points. My advice is to keep the number of questions around 10."

Key take-aways

- Document should be easy to fill out: "O" when OK and "X" when not OK
- Add comments only when there is an "X"
- Two types of questions
  - Housekeeping questions
  - Questions on Standardized Work observance
- Customize at your convenience. However, keep the number of questions around 10

FIGURE 8.2
Key take-aways on the auditing form.

My advice is to keep the number of questions around 10.

Thomas offered to summarize in a few points. He mentioned that the form should be easy to fill out. He commented, "Mark an 'O' when OK and 'X' when not OK. Add comments in reserved area only when there is an "X'" Also, he insisted that the number of questions, which are of two types, must be kept around 10 to keep auditing easy (Figure 8.2).

9

Sustaining and Auditing in Daily Management

Although he had said so previously, Daniel wanted to emphasize the importance of making auditing part of the daily management. He had too often seen initiatives being launched just to fade a few months or even just weeks later. He remembered the story that one of his friends, Bernard, who worked as a consultant told him a few months ago when they met by chance at Charles De Gaulle Airport. Bernard and Daniel befriended when they met several years ago in a plant of a famous car-component maker: Yakeo.[*] They were fresh graduates from French "Grandes Ecoles" and it was their first real professional experience, and their first contact with TPS[†] and Lean tools. They received their first Lean training in this company. Daniel still remembered that the company was the first French company to grab Lean tools and implement them in their plants on a large scale. In the early 1990s, the car-component company faced huge pressure from carmakers to reduce their prices. They had no choice but to adopt TPS very early. Their main motivation was cost reduction, which meant they used a "military" style with little care about the human dimension. Big pressure to reduce costs with little to zero human consideration led people to exasperation. Several people left, and the turnover skyrocketed. Daniel used to say, talking about the company, "I don't know anyone who currently works with Yakeo. The only people I know are former employees." Thomas and Bernard both agreed that they learned a lot during their tenure at Yakeo, but would never consider returning there at any salary.

Bernard and Thomas were recounting their memories of this period of their lives when Bernard interrupted Thomas:

---

[*] This name of the company is completely fictitious. Any resemblance is pure coincidence.
[†] Toyota Production System.

"You will never guess where I spent my last week of last year. I got a phone call from the director of the Yakeo plant where we started."

"Why did he call you?" Daniel interjected.

"Well, he wanted me to come over to carry out workshops to reintroduce Lean tools in his plant!"

"Wow! That's huge!"

"You bet, I could hardly imagine that I would have to go back to the plant where I learned Lean as a young engineer; I mean a plant that was so far ahead in the implementation of Lean at this time, to conduct workshop and train people."

"What happened to this plant?"

"I have no idea, but the result is there. Obviously, they have not been able to sustain and they slid back…."

"This is interesting, I see this happening a lot around me."

Daniel and Bernard then went to a nearby Starbucks café and finished their discussion around a cup of tea.

Daniel recounted the story to his colleagues. He wanted to call the participants' attention to two points. First, implementing a new system like Standardized Work might take some time, and a lot of effort but, as he put it, "this is actually the easiest part of the journey; sustaining is far more difficult." He also wanted to draw their attention to the importance of a balanced deployment of Lean in general. "If you only focus on what makes you save money and forget the human aspect and respect for people, you are building an unbalanced house that will ultimately break!" He shared some drawings with the participants that had two houses (Figure 9.1). The first house, which he called the Toyota house, had two pillars: "continuous improvement" and "respect for people." He commented, "This is how most Lean experts would characterize the Toyota house. Now, if you only focus on continuous improvement tools as I see more and more often, you will strengthen only one of the company's two pillars, and the weakest one, namely 'respect for people,' will break and bring your house down. Then you will have to rebuild the house again, just like in the story I told you about Yakeo."

Implementing a new system like Standardized Work might take some time, and a lot of effort but, this is actually the easiest part of the journey; sustaining is far more difficult.

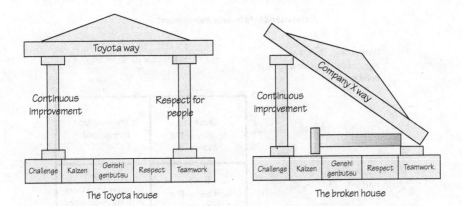

FIGURE 9.1
A single-minded focus on continuous improvement tools will lead to a "broken house."

If you only focus on what makes you save money and forget the human aspect, you are building an unbalanced house that will ultimately break!

Daniel then displayed a chart, which from his point of view should be used by managers for their daily management and problem-solving activities (Figure 9.2). "This is a chart that you can use any time there is a problem in the plant on a specific matter. Questions should be asked regarding standards. First of all, on the worker side, you want to answer the following question: 'Does he or she know and apply the standard on this matter?' There are two possibilities here: Yes or No. Now on the Management side, the key question to be asked is the following: 'Are there any standards on this matter?' As previously, there are two possibilities for answers: Yes or No. Now let us review all possible sets of answers, and decide what kind of decision can be inferred. In the first case, where there is a standard and the worker is trained, then you might need to review your standard. Chances are that it is not precise enough, or more generally was not done according to key rules we have learned throughout this week. In very simple terms: they are not real standards. Ask your best guy to lead the writing of the standards to replace the existing ones. In the second case, there are well-written standards but the worker is not trained. Well this is normal. It simply means that you have not been through the whole process of the establishment of Standardized Work.

Using standards in daily management

		Management side Are there any standards on this matter?	
When things do not go right, check the following points:		Yes	No
Worker side Does he or she know and apply the standard on this matter?	Yes	Check the quality of the standard!	Is the standard formalized?
	No	Train workers on the standard!	Establish Standardized Work with the support of managers!

FIGURE 9.2
Key questions to be asked when something goes wrong at a workstation.

The obvious response here is to train the worker using the TWI approach we just learnt. The third situation that might happen is when there are not standards and the worker has been trained to carry out his or her operation. This is a situation that at first look might seem awkward. The truth is that, it is actually the most common situation. In most of these cases someone who knows the job trains the worker. The trainer is preferably an experienced, or even the best performer. The point is that this person, who has reached the stage of unconscious competence[*] (Figure 9.3) does not always know what are the key points that make or break the job when it comes to productivity, quality, or safety. Lots of things are implicit, and besides that, the training method might not be the most efficient one. So at the end of the day, when you combine all those weaknesses, there is little surprise that people who are trained without using standards end up having lots of problems in their daily job. The way out is to actually write standards and use adequate documents—Job

---

[*] According to most experts, there are four stages of knowledge acquisition. In the first stage, the person is incompetent in a given subject but does not know it. This is called "unconscious incompetence." Then the person becomes aware of her or his incompetence. This is the stage of "conscious incompetence." From that point the person will seek to acquire skills until he or she becomes competent. This stage is called "conscious competence." The last stage comes when this person becomes and expert. Things then flow so naturally that he or she is not even conscious of his or her skills. This last stage is called "unconscious competence." Although they master their subject, people in this stage might not be the best trainers. They tend to see things that might appear complex to novices as trivial. Therefore, they fail to grab and explain key points to their trainees.

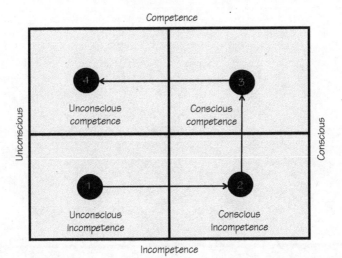

FIGURE 9.3
The four stages of knowledge acquisition.

Breakdown Sheets—to train workers. The fourth possible case is when there is neither a standard in place nor training implemented. Basically, nothing has been done. In this case, the whole process of Standardized Work implementation that we have been learning since the beginning of the week should be launched."

Daniel stopped a few seconds, then, pointing to the flipchart, he proceeded. "This chart is a synthetic roadmap that will help you move toward operational excellence (Figure 9.2). It's quite simple and it has lot of success with Plant directors. I strongly encourage you to always come back to the standards. In terms of manufacturing or process, this is your law or your bible. When it comes to operator's work, you should consider that any key point that is not in the standard does not exist. Again, it is clearly your internal law. This is how you will both sustain your hard work of writing standards, and harvest best practices and knowledge over the years." Daniel stopped and got a deep breath. Thomas jumped in and commented, "My understanding from what you just explained is that standards are completely integrated in production management. You do not produce parts on one side and sustain on the other one. The other point I liked from the story of Yakeo is the need to accompany continuous improvement with the human dimension, namely 'respect for people,' to avoid an unbalanced and thereafter broken house we saw previously" (Figure 9.1).

10

The End

The training was coming to its end. After a week spent together, participants were eager to take a well-deserved weekend. Most of them felt transformed by the amount of knowledge they had accumulated in only 5 days. Now their new mission would be to spread their learning in the plant to foster operational excellence. Thomas used an unusual phrase to describe their new role: "Lean evangelist." He asserted, "I want each of you to become Lean evangelist." The week had been exhausting for everyone, but all participants found reason for hope and even joy. There were some "bright spots," as Thomas said. Securing the support of the unions was a big win that would ease the implementation of the plant's competitiveness plan. He also recalled the first improvement results he celebrated yesterday. "I was happy to see some light in some participants' eyes yesterday, with the hidden hope that we will get more of this in the future," Thomas insisted. "The weeklong Standardized Work training has been transformative."

Before taking feedback from the participants, Daniel went back to the chart he presented the first day and checked the last two modules while explaining, "We have now covered every module of this chart (Figure 10.1). At this point, you have got everything you need to deploy Standardized Work as a system in your plant. Remember: 'system' is the key word. I hope you still remember that if you pick and choose one of the tools and deploy it alone,* you run the risk of a rapid relapse. It is now your responsibility to spread these tools in your plant."

A participant raised her hand to enquire about the second point of the last module on auditing: operator performance mapping (OPM). "Daniel, you have not said a word regarding this point. Could you tell us in few words what we need to know about that?" Daniel responded, "Well, there

---

* For more details refer to the first book of the series: *Implementing Standardized Work: Measuring Operators' Performance*.

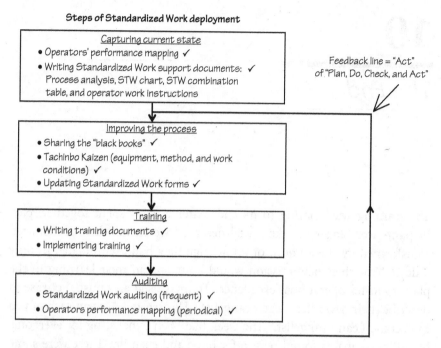

FIGURE 10.1

Complete training of Standardized Work.

is nothing more to say than what you learned during the first 2 days. You all know how to implement operators performance mapping. The only point about auditing is that you need to do this periodically. For instance, it could be every 6 months if your staff is stable. You can also perform an OPM following a big change in the worker organization. Just like a thermometer will help you measure the temperature, the OPM will give you a snapshot about the level of performance of workers. You will then be able to see how you are versus your target. I am sure you still remember the chart I showed you on day 2 of the training (Figure 10.2). There is nothing more to add to that." The questioner and the rest of the room seemed satisfied with Daniel's answer. In reality, most of them already had their head elsewhere and would not insist on anything that would defer the end of the training. After all, Daniel would still be accessible afterward. They could reach him virtually anytime if they needed to clarify any point of the training.

As the training was nearing its end, Daniel felt like the time was right to underscore some key messages before wrapping up. "As we saw earlier today, keep in mind that we are not only talking about continuous improvements, but people as well. So as you work to deploy Standardized

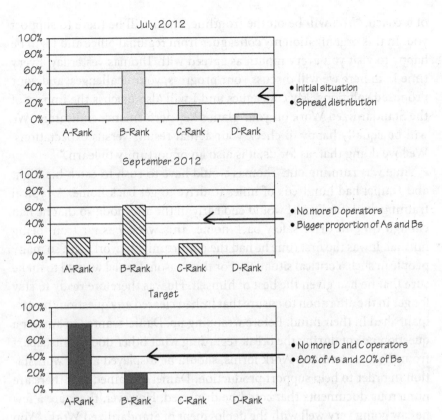

FIGURE 10.2
Periodical operator performance mapping shows progress toward the target.

Work, think about the human aspect. You need to bear in mind that change, even when it is for good, is not always welcome. This is especially true in this plant, where an important proportion of people have more than 20 years of seniority. Remember that they have been doing things a certain way for years, and now, all of the sudden, here are some folks showing up to ask them to work differently. As the saying goes: 'habits are second nature.' When you take some time to think about that, it is easy to understand their resistance. Managing change will be a big part of the work to implement. Do not underestimate the importance of it." Thomas jumped in and acquiesced. He insisted that he would be taking care of the point. Daniel also came back to one of the charts he had deployed on the first day.* He explained that his role will mostly become the one

---

* This figure is not needed for the understanding of the passage. However it could be found in the first book of the series: Implementing Standardized Work: Measuring Operators' Performance in Figure 4.3.

of a coach. "You will be on the frontline, and I will be there to support you. In this organization my colleagues from regional office and I will be happy to visit you every month as agreed with Thomas yesterday. Every time I am here we will discuss your progress, your challenges, and your proposed solutions. My colleagues and I will also discuss the impact of the Standardized Work on your plant's key performance indicators. We will be equally happy to check your actual results versus expectations. We love doing that, as for us, it is also an opportunity to learn."

Time was running out. Thomas would have to rush to catch his flight and Daniel had hundreds of miles to drive to get back home. A typical training week for Daniel would end early in the afternoon so that he had enough time to drive safely back home. This week was far from being normal. It was the first time he had the opportunity to work with so many people in such a critical situation. For this reason, Daniel wanted to make sure that he had given the best of himself. He was therefore ready to stay longer in the afternoon to ensure that he had cleared any questions participants had in their mind. Before wrapping up, Daniel wanted to answer a question he got during the break regarding what other documents, apart from the Standardized Work forms, should be displayed at the workstation in order to help support production. Daniel explained that there are numerous documents that could be displayed. However, there are a few he saw going very well with the deployment of Standardized Work. "You first want to make sure that everything related to safety is there. Look, when we talk about safety, one basic document that comes first is the one that shows the individual protection equipment (IPE) required at each workstation. Here is an example (Figure 10.3). Since safety and quality are 'must-haves' in any plant, good practice consists of displaying a big chart of the most important safety key point and the most important quality key point from the list you generated during the Standardized Work implementation. Now, how do you decide that a key point is important? Well, you may use one of the following two criteria: criticality or frequency. If there is an accident or a quality problem you never want to happen, then display it. This is what the first criteria is about. In some situations, you might observe that a safety hazard or a quality problem keeps appearing. It makes sense to display the key points that need to be observed to avoid them, thereby ensuring that workers are always reminded to be careful. On the chart over there, I displayed an example from the tire industry (Figure 10.4). For instance, the decision to display a key point on safety came from the observation that an increasing

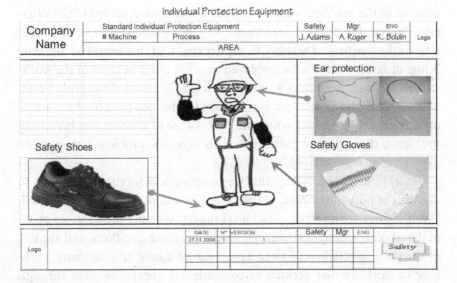

FIGURE 10.3
Mandatory individual protection equipment should be displayed at the workstation and made visible to everyone.

FIGURE 10.4
Key points regarding the most critical or frequent safety and quality problems should be displayed at the workstation.

number of tire builders were having back problems. Analysis of the situation showed that it was caused by the way they grabbed the tire to extract it from the machine. Therefore it was decided that this would become the 'king' of safety points to be displayed on a big chart in front of the workstation to raise awareness and keep it high. To conclude on this point, please note that if at some point you consider that progress has been made on a key point on display while the observance of another one is causing problems, then the key point related to this new problem becomes the new 'king' to be displayed."

Daniel punctuated, "Okay, this was a very good question. I planned to discuss the point but forgot, so thank you for asking." He then concluded, "Well, if there are no more questions, I suggest we enter the wrap-up phase of the whole training." Daniel answered a few more questions, and as most of the participants looked tired as a result of a long and intensive weeklong of work. Daniel decided to conclude. As always, he went through the whole room, head by head, to hear everyone's feedback. People were mostly thankful. They expressed their gratitude for what they had learned from the week training. Daniel was used to such an ending and would not give it more meaning than it carried. He also knew from his experience that the plant's journey was just starting. As he commented, "There will be lot of sweat, tears, and difficult times ahead, but if you are able to stay the course and pursue excellence relentlessly, you will prevail."

> *There will be lot of sweat, tears, and difficult times ahead, but if you are able to stay the course and pursuit excellence relentlessly, you will prevail.*

Index

Note: Page numbers followed by "*fn*" indicates "footnotes."

2 Sigma problem, 41

A

Auditing, 57; *see also* Daily management; Four-step method
    audit auditing, 63
    classical organization chart, 61
    classical pyramidal organizational chart, 59
    German culture, 66
    job position, 64
    key take-aways from, 65
    organization of Team leader auditing, 64
    PDCA, 58
    production management, 63
    reverse organizational chart, 62
    seven-member team organization, 60
Auditing document, 65
    audit sheet, 68
    comments in, 72
    key take-aways on, 71
    Standardized Work, 67, 69
    SWIP, 70

C

Classical organization chart, 61
Cloning spiral, 8
Company leadership, 17
Conscious competence, 76*fn*

D

Daily management; *see also* Auditing
    knowledge acquisition stages, 77
    and problem-solving activities, 75, 76
    single-minded focus, 75
    Standardized Work, 74
    sustaining and auditing in, 73
    TPS, 73
    trainer, 76–77
Double bonus, 26

E

End of Industry, 5
    cloning spiral, 8
    "end of Detroit," 6
    French industry, 7
    outsourcing, 7
    Standardized Work deployment, 9
Europe, Middle East, and Africa (EMEA), 1

F

Four-step method, 37; *see also* Auditing
    conservative estimation, 44
    face-to-face positioning, 38
    mastery learning, 40
    observation period, 44
    structured training, 43, 45
    studies, 42
    TWI requirements, 38, 40
    TWI's motto, 39
    2 Sigma problem, 41
    unstructured training, 43
Frequently committed mistake (FCM), 35

G

German culture, 6, 66

I

Individual protection equipment (IPE), 83

J

Job Breakdown Sheet (JBS), 3, 29, 30, 31

L

Lean, 17, 33, 60, 74
   deployment, 58, 74
   evangelist, 79
   training, 27, 73
Learning Pyramid, 27*fn*

M

Mastery learning, 40

O

Operator performance mapping (OPM), 79, 80
Operator Work Instructions, 3, 32, 33, 47–48

P

Plan, Do, Check, and Act (PDCA), 4, 19
Practicing training, 47
   digital material, 51
   for evaluation, 54
   evaluation grid, 53
   evaluation sheet, 55
   JBS, 49–50
   Operator Work Instructions, 47–48
   piece-by-piece training activities, 52
   Standardized Work combination table, 48
Preparation for training, 23
   breaking down T-shirt packing operation, 25
   double bonus, 26
   FCM, 35
   implementing Standardized Work, 24
   JBS, 29, 30, 31
   key take-aways from, 35
   learning and retention rates, 28
   Operator Work Instruction sample, 32, 33
   playing tasks for group cross training, 27
   T-shirt packing, 23
   time variations, 24
   training activities, 35
   TWI, 28, 29, 34
   variability, 26

Q

"Quiet brains," 57

R

Reverse organizational chart, 62

S

Situational Leadership, 40*fn*
Standardized Work implementation, 1–4, 13, 20
   auditing, 9, 11, 20, 57–66
   complete training, 80
   deployment, 2
   four-step method, 37–45
   implementation, 3, 4
   Job Breakdown Sheet, 3
   key points in auditing document, 67–72
   practicing training, 47–55
   steps, 14, 19
   sustaining and auditing in daily management, 73–77
   training, 9, 11, 23–36
Standard Work in Process (SWIP), 70
Structured training, 43

T

Time units (TU), 44
Toyota, 18, 53, 58, 69
   Toyota house, 74, 75
   Toyota Production System (TPS), 73
Training day
   auditing, 20

company leadership, 17
complete training of Standardized Work, 80
feedback from participants, 79
"Grandes Ecoles," 11
HR department, 11, 12
importance of training, 18
key messages, 80–81
key take-aways from training and auditing, 21
learning swimming, 16
mandatory individual protection equipment, 83
periodical operator performance mapping, 81
safety and quality problems at workstation, 83
Standardized Work Chart, 13
Standardized Work implementation, 14
successful training, 19
typical training week, 82–83
worker over time, 15
wrap-up phase of whole training, 84
Training within Industry (TWI), 3, 28, 34, 40
motto, 39
requirements, 38

U

Unconscious incompetence, 76*fn*
Unstructured training, 43

V

Variation, 24, 25, 26

Author

Dr. Alain Patchong is the co-founder and CEO of Mexence, an extension of the concept of his book series. Mexence is built around the simple idea that in today's highly competitive environment, industry, which has already harvested low-hanging fruits, cannot rely on single-minded or one-size-fits-all tools. Mexence partners with companies to improve their processes, to support their digital transformation and their automation process to deliver business excellence. Before co-founding Mexence, Patchong was the assembly director and master expert in assembly processes at Faurecia Automotive Seating. He was previously the industrial engineering manager for Europe, the Middle East, and Africa at Goodyear in Luxembourg. In this position, he developed training materials and led a successful initiative for the deployment of Standardized Work in several Goodyear plants. Before joining Goodyear, he worked with PSA Peugeot Citroën for 12 years, where he developed and implemented methods for manufacturing systems engineering and production line improvement. He also led Lean implementations within PSA weld factories.

Patchong currently teaches at Ecole Centrale Paris and Ecole Supérieure d'Electricité, two French elite engineering schools. He was a finalist of the Institute for Operations Research and Management Sciences (INFORMS) Edelman Competition in 2002 and a visiting scholar at Massachusetts Institute of Technology (MIT) in 2004.

He is the author of several articles published in renowned journals. His work has been used in engineering and business school courses around the world.